Springer Complexity

Springer Complexity is an interdisciplinary program publishing the best research and academic-level teaching on both fundamental and applied aspects of complex systems—cutting across all traditional disciplines of the natural and life sciences, engineering, economics, medicine, neuroscience, social and computer science.

Complex Systems are systems that comprise many interacting parts with the ability to generate a new quality of macroscopic collective behavior the manifestations of which are the spontaneous formation of distinctive temporal, spatial or functional structures. Models of such systems can be successfully mapped onto quite diverse "real-life" situations like the climate, the coherent emission of light from lasers, chemical reaction-diffusion systems, biological cellular networks, the dynamics of stock markets and of the internet, earthquake statistics and prediction, freeway traffic, the human brain, or the formation of opinions in social systems, to name just some of the popular applications.

Although their scope and methodologies overlap somewhat, one can distinguish the following main concepts and tools: self-organization, nonlinear dynamics, synergetics, turbulence, dynamical systems, catastrophes, instabilities, stochastic processes, chaos, graphs and networks, cellular automata, adaptive systems, genetic algorithms and computational intelligence.

The three major book publication platforms of the Springer Complexity program are the monograph series "Understanding Complex Systems" focusing on the various applications of complexity, the "Springer Series in Synergetics", which is devoted to the quantitative theoretical and methodological foundations, and the "Springer Briefs in Complexity" which are concise and topical working reports, case studies, surveys, essays and lecture notes of relevance to the field. In addition to the books in these two core series, the program also incorporates individual titles ranging from textbooks to major reference works.

Indexed by SCOPUS, INSPEC, zbMATH, SCImago.

Series Editors

Understanding Complex Systems

Founding Editor: Scott Kelso

Wei-Bin Zhang

Complexity Theory and Uncertainties

Interdependence Between Man, Society, and the Environment

Springer

Wei-Bin Zhang
College of International Management
Ritsumeikan Asia Pacific University
Beppu, Japan

ISSN 1860-0832 ISSN 1860-0840 (electronic)
Understanding Complex Systems
ISBN 978-3-031-42393-2 ISBN 978-3-031-42394-9 (eBook)
https://doi.org/10.1007/978-3-031-42394-9

This Springer imprint is published by the registered company Springer Nature Switzerland AG
The registered company address is: Gewerbestrasse 11, 6330 Cham, Switzerland

Paper in this product is recyclable.

Preface

We cannot solve our problems with the same thinking we used when we created them.

Albert Einstein

I first met with complexity theory in 1984 when I was sent by China's government, as one of made-in-China undergraduates, to Japan in 1983 as a graduate student in civil engineering at Kyoto University. Complexity theory—mainly Prigogin's and Haken's works—attracted me greatly (Haken, 1977, 1983; Prigogine, 1980, 1997). Four years later, I decided to purchase economics, applying complexity theory to economics. By the Spring of 1989, I completed *Synergetic Economics*, published in Prof. Haken's book Synergetics in 1991. This book is a continuation of my endeavor in complexity theory in man and society. It is a continuation and "updating" of my previous works. *Synergetic Economics* was, perhaps, the first comprehensive book on applying modern nonlinear theory and ideas from natural sciences to economics. It is a further development of Samuelson's *Foundation*. Since then, there are numerical publications on nonlinear economics. In recent years, I have further explored the implications of complexity theory for economics (Zhang, 2023a) and trust and justice in human evolution (Zhang, 2023b). This book gives a range of examples of recent advances not only in economics but also in other fields of man and society. Studying complexity theory requires some years of advanced mathematics to digest technical details. This book introduces some of the recent developments in applying complexity theory to man and society with minimum mathematics but plenty of simulation results with illustrative plots.

I am very grateful to Prof. Scott Kelso and Editor Dr. Johannes Glaeser for valuable comments, constructive suggestions, and effective cooperation. I am grateful to the projector coordinator Mr. Vijay Kumar for effective editing. I thank my wife, Gao Xiao, who is always supportive of my research. I completed this book at the Ritsumeikan Asia Pacific University.

Beppu, Japan
Summer 2023

Wei-Bin Zhang

References

Haken, H (1977) Synergetics: An introduction. Springer-Verlag, Berlin
Haken, H (1983) Advanced synergetics - Instability hierarchies of self-organizing systems and devices. Springer, Berlin
Prigogine, I (1980) From being to becoming. WH Freeman, San Francisco
Prigogine, I (1997) The end of certainty - Time, chaos, and the new laws of nature, written in collaboration with Stengers, I. The Free Press, New York
Zhang WB (2023a) Chaos, complexity, and nonlinear economic theory. World Scientific, Singapore
Zhang WB (2023b) Trust and justice: Complexity of man, complexity of society, and complexity theory. Lexington Books, Maryland

Contents

Chapter 1
Complexity Theory and Broadened Rationality on Man and Society

> The grand aim of all science is to cover the greatest number of empirical facts by logical deduction from the smallest number of hypotheses or axioms.
> Albert Einstein

Living within chaotic inflations, unprecedented pandemics, an international-scale war, environmental disasters, broken families, unforeseen population change patterns, and many uncertainties, one might reasonably feel about limited powers of scientific approaches to man and society. Human societies have largely failed in delivering happiness and welfare to people. Collectively, mankind is confronted with many natural as well as human-caused disasters. Natural laws, once constructed, remain invariantly valid in the sense that they can be repeated predictably in known conditions. Newton's laws are valid in a predictable way; so is Einstein's relativity theory. Water boils in the same predictable way, for instance, in Japan, Sweden, and Australia. Snow forms much in the same way under similar conditions all over the world. Disciplines such as astronomy, chemistry, physics, geology, and biology have developed a robust combination of logical coherence, causal description, explanation, and testability. These disciplines are mutually consistent. For instance, the laws of chemistry are compatible with the laws of physics, even though they are not reducible to them. Chemists do not propose theories that violate the elementary physics principles of the conservation of energy. Instead, they use the principle to make sound inferences about chemical processes. This property of the natural world enables man to effectively apply traditional scientific strategy which decomposes the whole into simpler parts until scientists can deal with simple parts.

Scholars who are concerned with man and society attempt to identify simplicity in a complex reality by the strategy widely and successfully applied in natural sciences. This approach has enabled various fields in academics to live in isolation from each other over quite a long period. Students trained in one subfield often do have not a shared understanding of the fundamentals of the others. When economists, for instance, from each subfield are busy with the pursuit of learning, they do not converge upon a common framework but find themselves in divergent directions. It is reasonable to observe that highly reputed economists in democratic societies would collectively give three different opinions—yes, negligible, or no—on any

W. Zhang, *Complexity Theory and Uncertainties*, Understanding Complex Systems,
https://doi.org/10.1007/978-3-031-42394-9_1

important political-economic issues of domestic affairs when they are faced with the same objective data. It has become evident that the traditional scientific method is not so effective in revealing the dynamics of man and society. Various mechanisms of conscience and non-conscience, being and non-being, emotion and rational calculation, egoism and empathy, physical and spiritual, inner world and environment, individual and family, and likely are simultaneously processing within the same body in harmonious, conflicting or unrelated relations between these forces. The *yin* and *yang* cannot be separated into different worlds because one does not exist without the other. Man has unlimited varieties of honesty versus manipulations conducted timely which are based on varied expectations of future consequences. One law for one society or in a special historical circumstance has little to do with general circumstances. Man is born to be creatively self-made, and society is always to be newly constructed.

The world is far better informed than at any time in history. But, the human capacity to predict behaviors of man, society, and the environment has not been much enhanced. There are more divorces after marriage vows and more unrealized dreams on personal levels. Climate changes are beyond comprehension. Prices often oscillate violently without any regularity. There are uncertainties in jobs and economic performances, exchange rates, and the like on national levels. In many parts of the world, people experience natural disasters in scale and scope that many regions never experienced. Aged people across the Earth complain that they have never experienced bizarre weather such as blazing heat, cold, frequent hurricanes, destructive Tsunamis, howling winds, pouring rains, and more. The world is full of uncertainties and chaos.

Traditional scientific sciences teach that if one has the right equations and sufficient data, phenomena should be predictable after considering various noises or external unpredictable shocks. A new science called chaos theory—exchangeably complexity theory, self-organization theory, nonlinear science, or synergetics in this book—demonstrates that even if one has sufficient data and the right equations, people can practically not predict what will happen in the future. Chaos theory mathematically proves that a nonlinear interdependence between two elements, such as *yin* and *yang*, or a man and a woman, can lead to unpredictable phenomena. Traditional sciences study only linear or linearized systems and treat nonlinear phenomena such as chaos and catastrophes as trivial or accidental. Recent developments in nonlinear mathematics and computers enable scientists to model, analyze, examine, follow, and see how nonlinear phenomena occur.

Humans live in the same world and observe the same phenomenon. But, people interpret the phenomenon differently in terms of origins and mechanisms. It was not a long time ago that a high proportion of the global population across the Earth started to understand and interpret social and natural phenomena within the same scientific framework. The dominant vision and framework of traditional sciences are Newtonian. If a mindset is framed in this vision, one would visualize that the movement of a system is predictable and will come to a stationary state or an equilibrium point if there is no external change in the environment and the time is allowed to last long enough. A typical case is the pendulum of a traditional clock. If the pendulum is moved away from its equilibrium, it will come back to its long-term equilibrium point without any more external force. Applying this idea to the economic system, for instance, Adam Smith argues that an economy that is composed some millions of households and workers will achieve a long equilibrium point if the system is dominated by a free market mechanism under law. There might be population growth and technological changes, but the consequences of these changes would be predictable and lead to a new equilibrium point with a single predetermined path of the adaption toward the new equilibrium point. This beautified description of market mechanism is still repeatedly taught in almost all fields of social sciences. Karl Marx has the same vision, but his equilibrium point is the collapse of capitalism owing to the inequality between the capitalist and the worker. But, historical paths of modern economies are characterized by multiple paths, bifurcations, chaos, and catastrophes rather than a prefixed point of collapse. Complexity theory rationally broadens traditional rationality in the sense that it concerns many phenomena, which are not dealt with by traditional rationality. It covers the territories of traditional sciences and goes far beyond them.

Complexity theory studies complex systems that are characterized by nonlinear interactions between many elements. It deals with how nonlinear interactions can bring about qualitatively new structures and how the whole is related to and different from its components. Research on complex systems has been sped up in association with the fast development of computers. A modern computer can explain a far wider class of phenomena than it could have been imagined even a few decades ago. The essential ideas about complexity have found wide applications among a wide range of scientific disciplines, including physics, biology, ecology, psychology, cognitive science, economics, and sociology. Many complex systems in those scientific areas have been found to share many common properties. The great variety of applied fields manifests a possibly unifying methodological factor in sciences. Complexity theory is bringing scientists closer as they explore common structures of different systems. It offers scientists a new tool for exploring and modeling the complexity of nature and society. The new techniques and concepts provide powerful methods for modeling and simulating trajectories of sudden and irreversible changes in social and natural systems.

Chinese intellectual tradition has faith, called Dao, which might, somehow, be the God by Spinoza, or truth. Chinese Daoism does not believe that a single and concrete faith can explain a variety of life. A minimum number of "forces" is two—called *yin* and *yang*. For instance, human nature is delineated by yin God (which is female, moon, water, evil…) and yang God (which is male, sun, fire, virtue, …). The competition of the two forces within one body gives rise to a variety of mindsets and behaviors. This is an "intuitive" vision or belief about man and society in East Asian civilizations before they were extensively affected by the West from the 1840s. The Chinese culture is characterized by applying this vision to almost all aspects of life. In philosophy, China has two pairs of Daoism and Confucianism: natural philosophy and humanism. Owing to the development of complexity theory, thinkers and scientists have an alternative vision of change. In Western, traditional dominant rationality is Newtonian; in East Asia is the yin–yang dynamics; and in nonlinear science,

(a) Newtonian pendulum (b) the yin-yang vision (c) complexity theory

Fig. 1.1 Three typical visions of changes. *Sources* **a** https://pixabay.com/zh/photos/subject-clock-pendulum-clock-2825833/ **b** https://upload.wikimedia.org/wikipedia/commons/thumb/a/a0/Yin_and_Yang **c** Ding et al. (2012)

both traditional Western and East Asian are special cases of nonlinear science, as illustrated in Fig. 1.1.

Newtonian determinism is illustrated by Laplace (1814):

> We may regard the present state of the universe as the effect of its past and the cause of its future. An intellect that at a certain moment would know all forces that set nature in motion, and all positions of all items of which nature is composed, if this intellect were also vast enough to submit these data to analysis, it would embrace in a single formula the movements of the greatest bodies of the universe and those of the tiniest atom; for such an intellect nothing would be uncertain and the future just like the past would be present before its eyes.

Rosen (1987) describes a complex system:

> ….a simple system is one to which a notion of state can be assigned one and for all, or more generally, one in which Aristotelian causal categories can be independently segregated from one another. Any system for which such a description cannot be provided I will call complex" (p. 324).

This hints that in a complex system, the causal categories become intertwined in such a way that it is impossible to find a dualistic language of state plus dynamic laws to describe it. Complex systems should thus be different from the so-once-claimed universal generalized dynamic systems. In complexity theory, one finds frequently used keywords such as catastrophes, bifurcations, trade cycles, chaos, pattern formation, the emergence of new structures, fast and slow processes, and the relationship between microscopic and macroscopic structures. All these topics cannot be effectively examined by traditional analytical methods which are concerned with linearity, stability, and static equilibria. Complexity theory has changed views about evolution. In complexity theory, a small change in, for instance, tax rates or exchange rates can generate unexpected results due to the complexity of networking and the sensitive dependent states. A tiny exogenous, perturbation, misunderstanding, or misinterpretation of the current state can be leveraged to catastrophic consequences.

A complex system may consist of several diverse elements combined in a fashion not immediately obvious. The various parts of a system may interact not only among themselves but also with the system, which in turn affects its constituent parts. The interactions of these elements may render the entire phenomenon more than the sum of its parts. This new theory examines the spontaneous formation of new macroscopic structures associated with discontinuous processes. It shows how parts of a system can organize themselves into new structures at certain instability points. The instabilities can form hierarchies leading to complicated patterns of structure and behavior. Descartes gives the principles and their manifestations: "Just as arithmetic consists of only four or five operations, namely, addition, subtraction, multiplication, division and the extraction of roots....so in geometry to find required lines it is merely necessary to add or subtract lines." As described by Ernst Mach, "Strange as it may sound, the power of mathematics rests on its evasion of all unnecessary thought and its wonderful saving of mental operations." I now use the most well-known model in complexity theory to illustrate why an interdependence between the two forces expressed in a simple mathematical equation can give rise to chaos.

A simple one-dimensional map is often used to generally illustrate insights into the idea of complexity resulting from simplicity. Consider the population and its change. Its values vary between 0 and 1–0 implying disappearance and 1 standing for the capacity that the environment can support the population size. We apply the logistic map to describe its motion. Let $x(t)$ stand for the population of society in period or generation t. If $x(t) = 1$, it means that the population achieves its capacity; if $x(t) = 0$, there are no people. The larger the value of $x(t)$ is, the society has more people. We interpret the variable $1-$ in t as the capacity left for sustaining more population. The logistic map, proposed by Robert May in 1976, shows the population t in period $t + 1$ is given as follows: $x(t + 1) = rx(t)[1 - x(t)]$. It should be noted that the Malthusian population growth model is given by: $x(t + 1) = (1 + r)x(t)$ because he does not take into account of constraints of resources on the population growth rate. For Malthus, the birth rate is largely determined by human natural desires. If one thinks of other factors such as resources, education, disasters, and wars, it is reasonable to consider that the growth rate of the population is not simply proportional to the past population.

In my interpretation, the coefficient, *r*, is a parameter. If one wants to make it more realistic, it is possible to treat r as nonlinearly dependent on other forces. I rewrite the logistic map as:

$$\frac{x(t+1)}{x(t)} = r[1 - x(t)].$$

I interpret this:

$$\frac{\text{The population of the next generation}}{\text{The population of the current generation}}$$

$$= \text{r x the capacity for population growth} \begin{cases} > 1 \\ = 1 \\ < 1 \end{cases}.$$

This implies:

The population of the next generation is possible

$$\begin{cases} > Thatofthecurrentgeneration \\ = Thatofthecurrentgeneration \\ < Thatofthecurrentgeneration. \end{cases}$$

This equation shows how the next generation's population is affected by the capacity for further population growth. In what degree the intergenerational relation is affected is determined by the propensity to have children and social, cultural, and political factors which are aggregately reflected in the value of r. For instance, r $= 2.5$ and $x(\text{thecurrent}) = 0.6$, then:

$$\frac{\text{The population of the next generation}}{0.6} = 2.5 \times 0.4 = 1.$$

This implies that Thepopulationofthenextgenerationis 0.6. The population is invariant over time in this special case.

Modern computer shows unexpected phenomena with this simple map. The map is a typical example of chaos theory. Simple interaction can lead to unlimited complicated phenomena as portrayed in Fig. 1.2. For instance, interactions between the population and the capacity for further population growth under physical, cultural, social, and economic conditions lead to unlimited behavioral patterns over generations. The dynamic force is simple—in the sense that almost any phenomena around us would be described by far more complicated dynamic systems—even though its hidden behavior could have only recently been revealed owing to recently available computers. The logistic map exhibits multiple types of long-term behavior that can be displayed by a dynamic system: stable and unstable fixed points, periodic orbits, aperiodic trajectories, and chaos. Here, random shocks are not required to produce chaos. One of the fundamental aspects of chaos is that possible motions are simultaneously present in the system. A particular manifestation of this is that there are typically an infinite number of unstable periodic orbits that coexist with the chaotic motion. Because such "chaotic" trajectories are unstable, errors of estimation in parameters or initial conditions, however tiny, will accumulate rapidly into substantial errors in the forecast. The characteristic of extreme sensibility to tiny perturbations is known as the butterfly effect. The future behavior of such a chaotic model solution cannot be anticipated from its patterns in the past. In our example, if r $= 2.6$, the society has a stable population pattern over history. In a simple agricultural society like Tokugawa Japan, the population exhibited stability

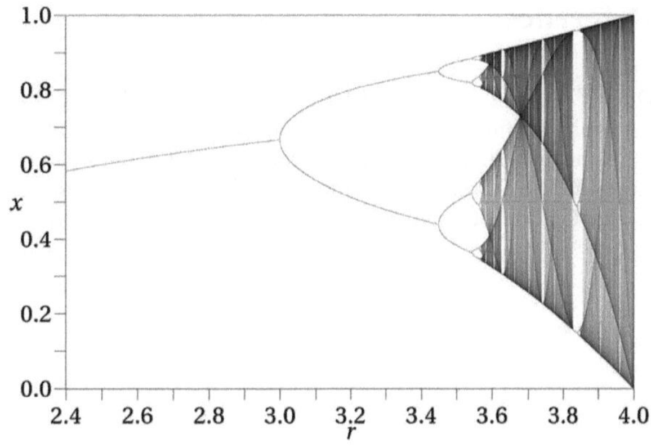

Fig. 1.2 Complexity of the population in different environments

over a long period. But when industrialization starts and urbanization occurs, population growth might not be formed in the same way as in a traditional society. For instance, the family might be more cautious about resource availability in deciding the number of children. We might have r = 3.2. Then a cycle repeats over generations. If society enters the service-, information-, and knowledge-based stage, then population dynamics are changed. When r = 3.5, the society experiences period 4 trust patterns. If r enters the chaotic zone, for instance, r = 3.8, the population exhibits chaotic behavior. It is impossible to predict population dynamics anymore as there is no regular pattern. Traditional theoretical studies on population cannot reveal the complexity of population dynamics with the simple growth mechanism. The model shows that even if one had a model, which perfectly describes the system, one needs precise measurements of initial states and parameters to know the future. Infinite precision is impossible in social and economic systems.

The *Yi Jing* emphasizes situation-dependent behavior: " 'The rain has fallen and is stayed': the power has accumulated to the full. 'If the superior man prosecutes his measures, there will be evil': he will find himself obstructed." The *Yi Jing* teaches that the essence of the world is dynamic. The *Yi Jing* holds that whether a society advances or becomes stationary also depends on the situation in linear progress as implied in Newtonian worldview: "We learn that advance will lead to difficulties while remaining stationary will afford ground for praise." There is a limit or boundary in movement. Any change is between the two extremes, the *yin* and *yang*. When the system arrives at either of the extreme poles, the system is not in a healthy (stable or humanistic) state. Although the actual state of the system will be located somewhere between the two poles, this does not mean that it is always possible to predict the state. It is only under special circumstances that the system is predictable. Change is not necessarily regular. Traditional Newtonian way of thinking is not invalid, but valid under limited circumstances.

References

Ding YT, Jiang WH, Wang HB (2012) Hopf-pitchfork bifurcation and periodic phenomena in nonlinear financial system with delay. Chaos, Solitons and Fractals: Nonlinear Science and Nonequilibrium and Complex Phenomena 45: 1048-57

Rosen R (1987) Some epistemological issues in physics and biology, in Hiley BJ, Peat FD, Quantum implications: Essays in honour of David Bohm, pp. 314–27. Routledge, London

Chapter 2
Influential Linearized Theories on Man and Society

Cohen and Steward (1995) argue:

> Mathematical descriptions of nature are not fundamental truths about the world, but models. … [W]hat model you use depends on the purposes for which you use it and the range of phenomena which you want to understand…. [A] degree of correspondence between deep underlying rules and reality that is never justified by any actual calculation or experiment (p. 410).

Modern thought on changes in man and society is mainly framed with the Newtonian description of linear mathematics. This vision of man and society is still framing the mindset of globalizing parts of the world. Cilliers (1998) describes the traditional rational approach:

> The traditional (or modern) way of confronting complexity was to find a secure point of reference that could serve as a foundation…. Whatever that point of reference might be…, following such a strategy constitutes an avoidance of complexity. The obsession to find one essential truth blinds us to the relational nature of complexity, and especially to the continuing shifting of these relationships (p.112).

It nowadays becomes evident that it is necessary to incorporate shifts and changes in different relationships in a structural sense when we try to comprehend complex systems such as man and society. Mouzelis (1995) illustrates:

> … rational choice theory tends to link micro with macro levels of analysis via logico-deductive methods that result in the neglect of 'emergent' phenomena and/or the various socio-historical contexts within which rationality takes its special forms. …[I]n so far as its mainly logico-deductive theorizing refuses to consider 'emergence', history, and context its statements (like all transhistorical, universalistic statements) tend to be either wrong or trivial (pp. 5-6).

For instance, as the formal economic theory of national development largely neglects laws and institutions, it provides little insight into important issues such as "Why did China achieve economic miracles?" It is illustrative to discuss a few influential ideas on man and society constructed with Newtonian vision.

W. Zhang, *Complexity Theory and Uncertainties*, Understanding Complex Systems,
https://doi.org/10.1007/978-3-031-42394-9_2

2.1 A Few Linearized Theories on Economic Evolution

> Simple can be harder than complex: You have to work hard to get your thinking clean to make it simple.
>
> Steve Jobs.

The traditional view of the relationship between laws and consequences — between cause and effect—holds that simple rules imply simple behavior; therefore, complicated behavior must arise from complicated rules. This vision has been recently challenged due to the development of complexity theory. Complexity theory shows how complicated behavior may arise from simple rules. A typical example is chaos identified from the simple logistic map, as shown before. In economics, for instance, there are many differences between traditional (or linearized) and nonlinear economics. Traditional economics has only three points of long-term economic evolution—destruction like Marx and Malthus, stationary equilibrium like Adam Smith, or unlimited growth like Schumpeter, as displayed in Fig. 2.1.

In socioeconomic dynamics, an attractor refers to properties of the variables and structures of the system which tends to evolve, regardless of the starting conditions of the system. The system moves toward this state space in a self-organizing manner. One may perceive an attraction. If one is attracted to something, one is drawn in by it. Smith and Jencks (2006) picture:

> Attractors, then. are not particularly strange. They are an adverse group of topographies. Mappings, routes, regularities, successful outcomes, capacities, knowledge forces and powers that pattern complexity. Conversely, they are negatively defined by no-go areas, obstacles, deserts, extravagant use of energy, propensity to catastrophic predation, poverty, ignorance, powerlessness, lack of freedom and opportunity (pp. 12-13).

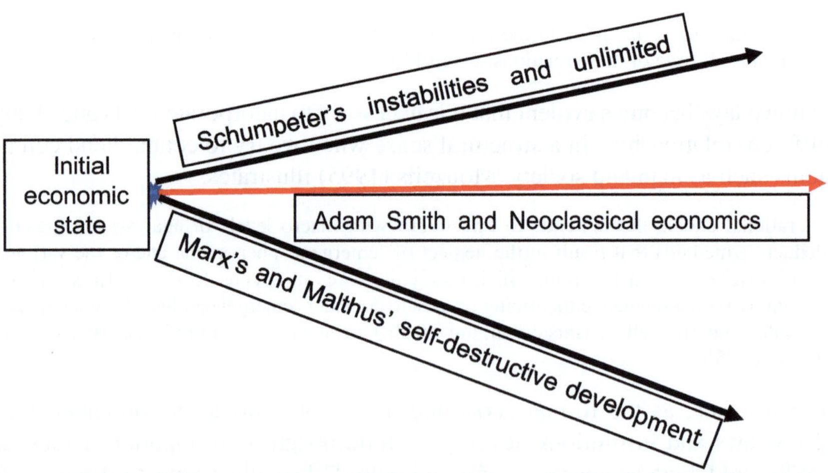

Fig. 2.1 The three possibilities of linearized economies

A fixed-point attractor, for instance, in Adam Smith's economic system, can be thought of as a balance of demand and supply of millions of people and firms who are motivated by self-interests, supported by a unique price (vector). Karl Marx does not believe in the existence of such an attractive attractor. He dreams up an economic system called communism which has an attractor characterized by a centralized planning system. Hayek and Mises publish extensively on the unsustainability of socialism. In contrast to an attractor, a repeller of a dynamic system is a set near which the flow is away from the set. The set of points is periodic or chaotic. For instance, Chairman Mao forced China into such a socioeconomic condition. Some small perturbations, for instance, by opening the country led the entire country far away from its poverty, low education, and inefficiency. A marriage based on pure romanticism may be a repeller, at least for some couples. A few quarrels break up the devout vow.

Adam Smith and the market mechanism

In 1776, Adam Smith (1723–1790) published *An Inquiry into the Nature and Causes of the Wealth of Nations*, which became the gospel of free trade and economic liberalism. Smith is regarded as the founder of a new science of political economy. He was the first to provide a systematic and comprehensive political-economic theory of free markets. He shows that economic agents driven by motives of self-interest make positive economic contributions to society. On the micro level, everyone is self-interested; but these self-centered agents produce the public good on the macro level. The economic miracle is achieved through an "invisible hand." The *Wealth of Nations* is the first book that so comprehensively and systematically treats economic principles and manifestations of these principles consistently. Smith constructs an economic theory of how markets operate under minimum government intervention. He analyzes economic principles of division of labor, determination of prices, rent, profits and wages, accumulation of capital, investment, distribution of capital and land among varied economic activities, regional as well as international trades, education, government policy, and many other issues. Smith's work has been the guide for successive generations of economists and the starting point of their speculation in theoretical construction.

In the *Wealth of Nations,* Adam Smith is concerned with the forces that govern the relative levels of prosperity among countries and that cause differences between countries. He regards the economic world as a vast workshop sustained by the division of labor with one psychological principle, the desire of everyone to better his lot. Different from Quesnay who pronounced the theory that agriculture is the source of all wealth, Smith held that labor is the true source of wealth. He discusses the advantages of the division of labor and its dependence on a scale of activity and the extent of the market. Large-scale activity and extended market permit specification, thus improving skills and labor efficiency. Technological progress, division of labor, and scale of the market are interdependent. The significance of free institutions under which people freely exploit the advantage of their skills, knowledge, and resources.

Adam Smith's economic theory with the division of labor and competitive equilibrium, irrespective of positive dynamic elements, sets limits on economic growth as diminishing returns ultimately prevail.

He was living in times of extremely unusual economic change—the stage of initial industrialization. This partly explains his relative neglect of physical capital accumulation and his conviction that agriculture is the principal source of Britain's wealth. He notes that industry generally affords greater scope for specialization than does agriculture and insisted that division of labor is necessary to increase wealth and that exchanges and the system of prices are necessary as the division of labor develops. His idea about economic equilibrium is proved by modern general equilibrium theory, while his theory on the division of labor is still in the process of improvement. He considers labor to be the sole standard of value but recognizes the existence of three factors of production: labor, capital, and land. Adam Smith recognizes the importance of exchange activities and the role of money. Division of labor cannot result in improved efficient reallocation of resources unless exchange activities are efficient. He argues how money could facilitate exchange by allowing people to avoid the difficulties of attaining the double coincidence of wants.

It is assumed by Adam Smith and mainstreams of economics that individual goal-optimizing behaviors, via free competition and cooperation (team formation), would collectively lead to strong and prosperous nations because individual successes add up the collective success. The whole is the sum of individuals. Nevertheless, complexity theory tells that the belief that the whole is equal to the sum of individuals is a special case, rather than a general law of evolution. The whole is larger or smaller than the sum of individuals. In economics, the whole is larger than the sum of individuals, in some narrow sense, when the increasing scale is dominant, and vice versa. Adam Smith considers possible increasing returns to scale, especially via learning by doing and market size. His well-cited pin factor story shows the idea that division of labor, market size, and improved productivity via learning by producing work together to sustain long-term economic growth. Market size is crucial for the division of labor which is a key factor in improving productivity for Adam Smith. Innovations and extensive research in industries by professions were not widely spread in his time. As workers repeatedly do the same task, their skills for the task will be enhanced. If the division of labor is extensively carried out in society, the productivity of the industry would be enhanced. The division of labor is thus an effect of economic growth and a cause of growth. The market size of the market is essential for the division of labor as workers cannot be specified in a single task if there is no sufficient demand. There are positive tradeoffs between growth, widened markets, division of labor, and learning by doing.

Schumpeter creates an unlimited growth world vision with increasing returns to scale, while Malthus depicts a dismal consequence of socioeconomic development with decreasing returns to scale. If society is composed of multiple sectors heterogeneously with increasing, decreasing, and constant scales, its complexity can, theoretically, lead to unexpected phenomena. Western civilization has been so successful in the last three centuries party because it has continuously created and explored economic activities with increasing returns to scale and exported activities with low increasing-returns-to-scale, constant, and decreasing scales to other parts of the world. This has resulted in global economic geography that poor countries have become poorer and rich economies have become richer.

Malthusian poverty trap and human disasters

Malthus (1766–1834), an English cleric, scholar, and influential economist, is best known for his "law of population." He was the sixth of seven children of his parents. He is the first to succeed in developing a theory of population growth. He published *An Essay on the Principle of Population* in 1798. He does not believe in the optimism of linear progress popular in his time. From 1798 to 1826, he published six more editions of his original text. Each edition updated the previous ones by incorporating new evidence and ideas. In his 1798 *An Essay on the Principle of Population*, he argues that an increase in a nation's food production improves the living conditions of the population, but the improvement is temporary because per capita food production is not enhanced, even though the national total output is increased. This implies that humans have the propensity to utilize resources for human reproduction rather than for maintaining a high standard of living. In 1798, he writes:

> Yet in all societies, even those that are most vicious, the tendency to a virtuous attachment is so strong that there is a constant effort towards an increase of population. This constant effort as constantly tends to subject the lower classes of the society to distress and to prevent any greater permanent amelioration of their condition.

The "virtuous attachment," the traditional marriage, would lead to constant efforts in producing more babies. For him, this is a natural tendency. By emphasizing the interdependence of population growth and food supply, his theory supports the subsistence theory of wages which has important influences on Ricardo, Marx, and Keynes. According to Malthus, the population tends to grow, and the lower class will eventually suffer hardship, want, and even famines and diseases. In his 1798 book, he explains:

> If the subsistence for man that the earth affords was to be increased every twenty-five years by a quantity equal to what the whole world at present produces, this would allow the power of production in the earth to be absolutely unlimited, and its ratio of increase much greater than we can conceive that any possible exertions of mankind could make it... yet still the power of population being a power of a superior order, the increase of the human species can only be kept commensurate to the increase of the means of subsistence by the constant operation of the strong law of necessity acting as a check upon the greater power.

By explaining poverty in terms of a simple interdependence between the population and the means of subsistence, the theory still provides important suggestions for poor countries' economic policies. Malthus' economics is one of the first economies functioning with non-constant returns to scale. It is decreasing return to scale due to population growth that drives the economic system into miserable situations. He also analyzes the possibility of social progress by pointing out various checking forces of population and poverty.

Malthus is aware of uncertainties in economic evolution with endogenous population. Malthus (1817) emphasizes irregularities in economic development: "A faithful history … would probably prove the existence of retrograde and progressive movements … the times of their vibrations must necessarily be rendered irregular" (p. 91). He believes in the non-linearity of structural relations, and their complicated multi-connected nature. For him, there are positive and negative factors for population dynamics. Population growth is checked by rises in the death rate, which can be caused by hunger, disease, and war. Factors such as celibacy, birth control, and postponement of marriages can also slow down population growth. He argues that if society limits population growth by human misery such as famines and wars, economic cycles may appear. He also points out that if society encourages, for instance, later marriages, population growth may be reduced and living standards are enhanced. This is obvious in the case of China where birth control was once associated with rapid economic growth. In his approach, it is important to analyze the properties of a dynamic system "out-of-equilibrium" and to trace the paths by which equilibrium might be approached. His conclusion on the relationship between population and economic growth is not applicable, for instance, to developed economies, but still provides important insights into the population growth of poor economies. One obvious obstacle for him to correctly explain modern economic growth is owing to his limited perceptions about technological changes, marriage markets, female labor participation, and other modern economic phenomena.

Karl Marx, inequality, and collapse of capitalism

Central issues to Marx's economics are issues related to the fundamental determinants of relative processes of production and reproduction and the distribution of income among social classes. The idea of specialization by division of labor provides guidelines for exploiting the comparative advantage principle from the individual factory floor to the industrial corporation as a home. The more limited the number of tasks performed by each worker the more comparative advantages would be used and the higher productivity and profitability of production would be. Marx foresees some of the political consequences of the unrestricted division of labor. He holds that in the long run, nobody would be able to set up competitive industries owing to the complete fragmentation of the production process. Two classes would emerge—the fragmented and specialized labor class and the capitalist class.

Marx considers the distribution of income between wages and other incomes as the key to explaining the processes of capitalist systems. His economics assumes that income distribution is determined according to two classes, capitalists, and workers. Class division plays a key role in the determination of real wages and the determination of capital accumulation. Group power determines the real wage rate. Capitalists are assumed to have a higher saving propensity than workers. In his analysis of capitalism, Marx neglects the possibility of class transformation and the government's positive intervention. For instance, by providing educational opportunities to children of the working classes, the government may make class transformation possible in the long term. If endogenous class transformation is considered in an industrialized capitalist society within the framework of endogenous knowledge and capital accumulation, Marx's conclusion about capitalism may be invalid in a long-run perspective.

Extreme inequality implies that the poor have little access to healthcare, educational conditions, and social mobility. The promise that the poor can improve their living conditions by diligence may not be realized in capitalism if there is no government intervention. Diligence is often sustained by foreseeable opportunities rather than wasting time, limited resources, and passions. The difference between the rich and the poor is becoming enlarged in modern societies. A rich family has wealth for many generations to consume, while a poor family has no food for tomorrow. It would be difficult to conceive how a young man born into a poor family in a prosperous and democratic society can establish a realizable dream for a sustainable happy life. Born into poverty implies that in the entire life, one hears about beautiful promises from society and watches all the lifestyles of the rich and the beautiful daily till the end of life without the ability to even enjoy a stable and proper family life. Enlarged inequality and difficulties of upward social mobility don't create the dreams of those born into poverty. For Marx, extreme inequality destabilizes society.

2.2 John Rawls' and Piketty's Linearized Approaches to Justice

Bertrand Russell comments on certainty: "The demand for certainty is natural to man but is nevertheless an intellectual vice." Justice is the basis of a desirable society. As globalization is deepened, global justice, and local justice are not necessarily in harmony. One of the influential modern thinkers on the theory of justice is John Rawls (1921–2002). He was born in Baltimore. He graduated from Princeton University in 1943 with a Bachelor of Arts. He had a stutter and a limelight problem. He received his PhD from Princeton in 1950 with the dissertation "*A Study in the Grounds of Ethical Knowledge: Considered concerning Judgements on the Moral Worth of Character*." He became a moral and political philosopher. He received the National Humanities Medal in 1999 presented by the president of the United States for his work, which "revived the disciplines of political and ethical philosophy with his argument that a society in which the most fortunate help the least fortunate is not only a moral society but a logical one."

His 1971 book *A Theory of Justice* caused a great interest in normative political philosophy in academic cycles. He is still recognized as one of the most influential philosophers of the twentieth century. Rawls (1971: 516–517) states America's need for relating justice with public reasoning for justifying the American conception of human rights in the gradually connected world:

> We don't look at the social order from our situation but take up a point of view that everyone can adopt on an equal footing. In this sense we look at society and our place objectively; we share a common standpoint along with others and do not make out judgments from a personal slant.

He builds a theory of social justice by designing the original position in which people intellectually and rationally select the kind of society they would choose to live in. He constructs his theory via a reflective equilibrium method. It is assumed that people know which social position they would personally occupy. People are assumed to be ignorant about their distinctive traits, such as intelligence, skills, wealth, gender, race, age, education, and even religion. The only thing that people know about is some basic capacities necessary to participate in a sustainable system characterized by cooperation freely and happily. A person has two basic capacities. One is that everyone can form, pursue, and adapt a desired lifetime pattern or plan for a good life. There is no need for any person to know what the good life is in the initial position of ignorance. The other one implies that any person comprehends that he can develop a sense of justice and a stable desire to abide by it. Under these conditions, any person would improve the position of the worst off, because one might find oneself in that position. Sen (2006) challenges Rawls' theory from different perspectives. In the light of complexity theory, his visionary framework can hardly provide any insight into the complexity of American, not to say global, civilization, because there is not such an equilibrium even in very ideal conditions in the long term. His imagination might come from the early American conditions that a few people came to a vast "empty" great land and built a country that did not have many conflicts of interest, at least for white people. When resources—such as high social status and positions in professionals—are more competitive and manipulations and deception are common, his theory has little ground for reality.

His theory constructs a liberal society with freedom and equality. Rawls contemplates justice as fairness. Basically, like Adam Smith, Rawls believes that he could solve natural conflicts of people with various interests through his political construction. He argues for equal basic liberties and equality of opportunity. Society should maximize the benefit to the least advanced members of society in any case where inequalities may occur. He believes that freedom and equality can be integrated into a harmonious society in the long term. Indeed, he does not have a contemporary conception of stability and complexity. Rawls argues that under his assumed world, people will choose fair policies. The liberty principle is maintained in society in the sense that everyone will enjoy basic liberties, which refer to freedom of conscience, democratic rights, property rights, and association and expression.

As there is no evil nature of men, such as deception, cheating, or natural propensity to break laws, in his ideal world, his theory is misleading if (some) men are smart, have access to education and other equalized opportunities, and are evil. The construction of simple gaming with learning processes shows that the inner evil (smart) men (and shameless women) have a high probability to dominate a democratic society. As he has little idea about the endogenous accumulation of capital and knowledge, he could not image great inequality under liberty and free markets, as a contemporary undergraduate in economics knows. As work opportunities are controlled by capitalists, politicians, and elites in free societies, plenty of people born into poverty have little opportunity to enter high societies before they lose the spirit of freedom. Human history does not show any sign of such a society that satisfies one of his principles. Sen (2006) comments: "We have to move the theory of justice out of that little room" (p.238). Complexity theory mathematically confirms that a complex system cannot avoid chaos even if it is well-designed. In capitalist systems, a difference in inequality can be rapidly enlarged and exaggerated over time. One day, one has an ideal situation. After a few years, all the initial desirables or undesirables may be turned into their opposites. Plato hopes: "Our object in the construction of the state is the greatest happiness of the whole, and not that of any one class." *A Theory of Justice* reads like a process of justice that a group of people who come to "an empty" land to build a prosperous country.

Plato refers to injustice: "When a Benefit is wrongly conferred, the author of the Benefit may often be said to injure." Adam Smith argues that specification and division of labor would increase the efficiency of economic production. Karl Marx is concerned with a society with two classes—one has the wealth to earn profit and the other one has no wealth but labor to sell. Keynes advocates strong government intervention in capitalist economies. Under Keynesian policies, capitalist economies have enjoyed prosperity and socially tolerable inequalities for quite a long period. Nevertheless, the basic issues that Marx addresses do not disappear and are increasingly causing more social concerns and conflicts.

When discussing justice, I should also mention another influential scholar on justice and inequality. In 2014, French economist Thomas Piketty (1971–) published an influential book, *Capital in the Twenty-First Century,* causing attention to the issues from academics as well as the public. He studies wealth concentration and distribution over the past 250 years. His approach is Newtonian—identifying some single force to explain complexity. All complexity is reduced by a simple and clear law of enlarging inequality: the rate of interest > the growth rate. Piketty (2014) argues: "Social inequalities are acceptable only if they are in the interest of all and particular of the most disadvantaged social groups" (p. 480). His framework is a combination of Marx's approach and the Solow model. Like Marxian economics, he divides the population into two classes.

I now illustrate his basic idea. The capital share of the national income is $(r \times K)/Y$, where r is the net real rate of return on capital, K is the capital stock, and Y is the national income. The total return from capital is rK. For a given level of the national income, the larger this number becomes, the more income the capitalist will get. Piketty introduces the fundamental law: $K/Y = s/g$, where s is the saving rate and g is the growth rate of the national income. This relation is directly from the Solow model. From the two equations, one has:

$$\text{Capitalshareofnationalincome} = (r \times s)/g.$$

Piketty presumes that the growth rate varies, but the interest rate and saving rate are relatively stable. Accordingly, a lower growth rate leads to a higher capital share. He argues that if $r > g$, inequality tends to be enlarged. This occurs because capital income tends to increase at the interest rate and national income increases the growth rate. Piketty (2014) describes the implications of $r > g$:

> This fundamental inequality will play a crucial role in this book. In a sense, it sums up the overall logic of my conclusions. When the rate of return on capital significantly exceeds the growth rate of the economy, then it logically follows that inherited wealth grows faster than output and income" (pp. 25-26).

Piketty (2014) explains the inequality before WWI:

> The primary reason for the hyper-concentration of wealth in traditional agrarian societies and to a large extent in all societies before World War I is that these were low-growth societies in which the return on capital was markedly and durably higher than the rate of growth (p. 351).

He also predicts what will happen in the future: a strong tendency to have a larger inequality. Acemoglu and Robinson (2015, see also Zhang 2006, 2020a for basic theories) extensively, both theoretically and empirically, point out the limitations and invalidities of Piketty's approach. Acemoglu and Robinson (2015) conclude:

> argued that Piketty goes wrong for exactly the same reasons that Karl Marx and before him David Ricardo, went astray. These quests for general laws ignore both institutions and politics, and the flexible and multifaceted nature of technology, which make the responses to the same stimuli conditional on historical, political, institutional, and contingent aspects of the society and the epoch, vitiating the foundations of theories seeking fundamental, general laws (pp. 24-25).

In his recent *Capital and Ideology* (Piketty 2020), he advocates for a new "participatory" socialism, which is characterized by an ideology of equality, social property, education, and shared knowledge and power. Since his way of reasoning is Newtonian, a clear and logical style with an explicit and unique conclusion is expected.

2.3 Fukuyama's Linearized Approach to Trust

Justice connects all levels of human interactions over time across the Earth. Trust is often local, and trust based on fairness is not easy. To talk globally and to act locally reflects possible conflicts between trust and justice. Gangsters might trust each other far more than contemporary politicians in democratic societies, even though none truly cares about social justice in depth. Gangsters might even have a more proper sense of justice than politicians. In no society, politics is run by decent politicians and society is full of gangsters. Gangsters exist because of corrupt politicians and irresponsible citizens. There is trust between informal gangsters and formal governmental officials. Ancient Greek philosophers justify slavery but emphasizes trust within the localized group. Confucius' comment that a son in some village might not reveal his father's wrongdoing of stealing a neighbor's sheep is still causing hot discussions among moral thinkers. Harmony or conflicts between trust and justice are often determined by the boundary between the inside and the outside. But the boundary is chaotic. For instance, the definition of "citizen" who has the right to vote changed in the last two hundred years in America.

Fukuyama (1952–) caused a great interest in discussing the role of trust in socioeconomic development. He argues, in the late 1990s, that trust is vital in the future because network plays a key role in business. Bruce Arians (1952-), an American football coach, values the importance of trust:

> Familial betrayal is, to me, the most heartbreaking kind- because if you can't trust your family to love you and protect you. ... Trust brings a higher level of communication and a higher level of commitment and accountability.

The place where the most heartbreaking butyral is often the same as in the field where trustfulness is highly valued and trust is seemingly solid, as in Germany and Japan. In China, it often occurs between relatives and friends. Deceivers know the weakest point to get the demand satisfied.

Fukuyama (1995) published a well-accepted book, *Trust: The Social Virtues and the Creation of Prosperity*, on issues related to trust and economic growth. The book is framed in a traditional Newtonian approach. A single variable, trust, is used to explore the complexity of market economies. Trust has a complicated interdependent relationship with many other variables. The culture of trust is treated as the source of spontaneous sociability. By comparing different societies, he groups societies into high-trust societies such as America, Germany, and Japan, and low-trust societies such as Italy, France, Korea, and Taiwan. The four low-trust societies are called familistic societies in which the family is the basic unit of economic organization. These low-trust economies had difficulties in organizing large organizations. For him, Japan and Germany have high trust levels. It was thus quite easy for the two economies to create large companies with modern professional management and trust workplace relations on the factory floor. He believes that ethical habits are important for organizational innovation and the creation of wealth. His work is often criticized because he does not pay much attention to the distrust aspects of Japanese groupism. He admires Japanese *keiretsu* as a symbol of high trust and sociability. Japan was in its golden time and trust in the government was high when Fukuyama was conceiving of Japanese society. Since the late 1990s, trust in Japanese society has experienced a crisis (e.g., Kikuchi 2018). The whole society has been experiencing an almost linear decline of trust in almost all aspects over the last three decades. The Japanese public has been frustrated constantly by cheating and fabrications of companies.

Fukuyama (1995) holds: "Trust arises when a community shares a set of moral values in such a way as to create expectations of regular and honest behavior." In Japanese companies, it is not honest behavior but a secure position that counts most. Napoleon gives a law of social and economic exchanges: "The surest way to remain poor is to be an honest man." What Napoleon—a self-confident, powerful, romantic, experienced, deceptive, and successful man—says is well observed over history. Behind trust, there may be hidden distrust. A great society, like a popular man or powerful beauty, is not characterized by the so-called trust as Fukuyama calls. A society's appealingly most trusted state may be experiencing the most rapid deterioration of trustfulness. Fukuyama emphasizes the emotional cultivation of civil society. Nevertheless, civil society may be a collection of unproductive people or even a consumer club whose net contribution to society is like a parasite in the social organ. The development of civil society dramatically increases the complexity of social networks. Children in metropolitan areas are often taught to trust no stranger and almost all people with whom these children meet are strangers. Even within the family, the husband may not trust the wife and vice versa.

A half-truth is a whole lie, and two lies have nothing to do with truth. The master of lying benefits from society. The father, the son. Social scientists and economists who have followed the Newtonian approach tend to make similar mistakes when they explain social and economic phenomena, picking up one aspect to explain everything. Trust and justice are paramount for effectively organizing and managing family, kinship, private business, voluntary associations, professional associations, state enterprises, political parties, religious groups, and the state. There does not exist a linear relationship between trust and economic efficiency (e.g., Tabellini 2008, 2010; Marini 2016). The more complicated a society is becoming, the more convoluted "classifications of trust" will become. "Trust no one but yourself," is said everywhere. According to Daoism, Fukuyama's trusted society should be full of hypocrites, great thieves, and criminals. America is such a place, even though one might wonder why American Japanese Fukuyama could not recognize this plain fact. Effective deception takes place not so often among mutually distrusted people as among mutually trusted people. Trust is maintained for untrustfulness. Niccolo Machiavelli reveals: "A man who wants to act virtuously in every way necessarily comes to grief among so many who are not virtuous. … It is a double pleasure to deceive the deceiver." In China, when one officer is exposed publicly for corruption, there is usually a group of officers involved. Frauds, cheating, criminals, and corruption in large organizations are usually conducted not by a single individual, but through cooperations with professionals who appear to be innocent and claim to have been "blind" to wrongdoings around in the name of professionalism.

2.4 Politics in the Division of Labor

Ronald Reagan describes: "It has been said that politics is the second oldest profession. I have learned that it bears a striking resemblance to the first." Politicians and prostitutes are the most ancient professions in human societies. Both play changing roles in solving the creation, distribution, and utilization of wealth, power, and sex over history. Manipulation is a natural part of emotional play. Both politics and prostitution play human emotions. In the two oldest professions, there is one for which no professionalism is required because the masses cannot evaluate the performance properly. But one lies optimally in darkness, while the other lies in public. Politics is the only—perhaps similar to gambling in modern economies—important profession in that politicians have little idea about what they are doing and cannot even know the consequences of their decisions. President Reagan, who does not seem to master advanced academic knowledge in any field but was skillful at playing sports and human emotions played wonderfully politically in front of the masses and on screens. Michael Laver describes:

> Politics is about the characteristic blend of conflict and co-operation that can be found so often in human interactions. Pure conflict is war. Pure co-operation is true love. Politics is a mixture of both (Cees 2018).

Politic activities can be violent and non-violent. Mahatma Gandhi argues: "It is better to be violent if there is violence in our hearts than to put on the cloak of nonviolence to cover impotence." There are some well-known quotes on politics:

> Politics have no relation to morality—Nicolo Machiavelli
>
> The ballot is stronger than the bullet. - Abraham Lincoln.
>
> It is enough that the people know there was an election. The people who cast the votes decide nothing. —Joseph Stalin
>
> Politics is war without bloodshed while war is politics with bloodshed. —Mao Zedong
>
> The revolution is a dictatorship of the exploited against the exploiters. —Fidel Castro
>
> The modern conservative is engaged in one of man's oldest exercises in moral philosophy, that is, the search for a superior moral justification for selfishness. —John Kenneth Galbraith
>
> In politics, stupidity is not a handicap. —Napoleon Bonaparte

Society is characterized by the division of labor and the division of consumption. Politics is essential for social order. It is a part of society. Rules. institutions, the structure of social status, and faiths are integrated parts of social and economic evolution. They are often decided and performed by politicians in the division of labor. Politics penetrates every part of society, like blood in the human body. An inefficient and corrupt political culture implies the dysfunctions of society. Complexity theory tells that when affairs are interdependent, it is often difficult to decide the causes and the effects, even though theories often assume what determines what. Portuguese writer Jose Saramago (1922–2010) reveals: "It is economic power that determines political power, and the government becomes the political functionaries of economic power." This is not universally applicable. In China, for instance, it is the guns that Mao "successfully" managed China until his death. In capitalist systems with functional law, wealth functions far more freely than political power as money has no name, no identity, and sometimes even no trace, while political power is subject to legal and moral control. In capitalist systems, money penetrates everywhere either in darkness or in public. Politics cannot control capitalist money, but capitalist money achieves its purposes with an invisible hand.

Democracy is a symbol of political justification in many countries. Even the government of mainland China, which has a global reputation of dictatorship and brutality against those who criticize its authority, claims itself democratic. Democracy is not a unique way to build a decent and effective government. It is essentially a process of establishing authority via the election of those who are entitled to vote. Democracy cannot guarantee that there is an effective process in selecting capable leadership because the masses often fail to collectively choose capable leaders. George Orwell observes: "In our age there is no such thing as 'keeping out of politics.' All issues are political issues, and politics itself is a mass of lies, evasions, folly, hatred and schizophrenia." "The more I see of the representatives of the people, the more I admire my dogs." said Alphonse de Lamartine. Plato says: "Opinion is the medium between knowledge and ignorance." Masses in democratic societies have a lot of opinions inflated with fake news.

American economist Thomas Sowell observes the behavior of the American government:

> In various countries and times, leaders of groups that lagged, economically and educationally, have taught their followers to blame all their problems on other people – and to hate those other people.

Since effective democracy requires properly informed, well-educated, and non-partisan citizens, few prosperous, not to say poor or undeveloped, countries can be claimed decently democratic. The government can seldom correctly justify what it is doing if politicians in the political system are not even as honest as common humans. Lying, manipulations, no faith, and no principles are the main features of some politicians. The political system is only one complexity of social complexities. A dirty heart can bolster a healthy body. Similarly, dirty politicians can maintain a prosperous and well-mannered society. But politicians in gaming, like other disciplines, need to be honest, disciplined, dutiful, and professionally loyal. John Stuart Mill argues:

> The person who has nothing for which he is willing to fight, nothing is more important than his own personal safety, is a miserable creature and has no chance of being free unless made and kept so by the exertions of better men than himself.

In politics, one may reasonably doubt the possibility of being associated with better men.

It is well mentioned that Wall Street is the accumulation center of both American wealth and political power. Although the former is measurable to some degree, the latter is largely intangible. It does seem acceptable to consider that political power is manipulated and even controlled by the rich in capitalist societies. Once the political system cannot be effectively separated from the economic system, interactions between the two systems can lead to unlimited behavioral consequences. Thurgood Marshall (1908–1993), describes America: "Our whole constitutional heritage rebels at the thought of giving the government the power to control men's minds." Politics is a world run by different mechanisms. As political decisions relate various functions of society together, it is often not solid reasoning but opinions or simple emotions that characterize politics. Geroge MacDonald sees: "It is not in the nature of politics that the best men should be elected. The best men do not want to govern their fellowmen." A wise man does not speak affairs that he cannot comprehend. For Ernest Benn, "Politics is the art of looking for trouble, finding it everywhere, diagnosing it incorrectly and applying the wrong remedies."

Plato dreams about an ideal state of human society:

> There will be no end to the troubles of states, or of humanity itself, till philosophers become kings in this world, or till those we now call kings and rulers really and truly become philosophers, and political power and philosophy thus come into the same hands.

Many rulers of the world have been daily supplying the stock of humor and mocking over billions of screens connected by digital technologies because of their stupidity, dishonesty, cunningness, bad manners, boring and routine speeches, ignorance, biased opinions, narrow minds, being partisan, and the like. Anyone who thinks seriously and observes objectively the human quality of politicians across the Earth reasonably doubts how far mankind has been away from monkey, wolf, and ant societies. If justice and morality are used to distinguish animal societies from human ones, one might ask in what measurements man has made meaningful progress in politics about knowledge, justice, and fairness in association with fast technological changes.

Education and objective knowledge have been enhanced in most professionals, except politicians. Politicians still similarly govern as they did control and manipulate the world people of ignorance and illiterate. Many parts of the world are dominated by corrupt businesses and irresponsible politics. Bertrand Russell (Nissani 1992) observes: "It is the custom among those who are called 'practical' men to condemn any man capable of a wide survey as a visionary: no man is thought worthy of a voice in politics unless he ignores … nine-tenths of the most important relevant facts." Napoleon Bonaparte, who is reputed for political and military genius, honestly comments: "In politics stupidity is not a handicap." Thomas Sowell explains the division of the economist and the politician in managing the state: "The first lesson of economics is scarcity: there is never enough of anything to fully satisfy all those who want it. The first lesson of politics is to disregard the first lesson of economics." "It is always easy, glorious, and even human to spend money of other people for one's own or organization's benefits." Obedient economists justify any state intervention of the government because they are servants to the powerful.

Complexity implies multiplicity and situational dependency. One problem has multiple—sometimes seemingly rational and optimal—solutions. Politicians can apply their manipulating skills—which they never lack because manipulation is what they have learned in political or other professional careers—and massage the mass' emotions and keep critical intellectuals to be silent or neglected by influential media. The main voices of societies are often controlled by capitalists who have unbreakable relations with political systems. Politicians are often from and work for the benefit of the rich in the long term. Common people are daily watching politicians who are shaming their nations and races owing to their lack of cultivation, ignorance about contemporary reality, and unfair judgment. People often complain about their politicians, but rarely about themselves. It is their irresponsibility for society, blind obedience, and no learning that stupidity or dictatorship spread across society. Complexity theory tells us that it is not only the politician but the collectiveness of the mass mindsets that determine and allow the inefficiency of politicians in a certain cultural environment.

References

Cohen, J, Stewart M (1995) The collapse of chaos, Penguin, Harmondsworth.

Cilliers P (1998) Complexity and postmodernism: Understanding complex systems. Routledge, London.

Mouzelis N (1995) Sociological theory: What went wrong? Routledge, London.

Smith J, Jencks C (2006) Qualitative complexity. Routledge, London.
Malthus, T. (1817). Principle of population, 5th ed., with introduction by Blaug M, 1963. Richard D. Irwin, Homewood, IL.
Rawls J (1971) A theory of justice. Belknap Press, Cambridge
Sen A (2006) What do we want from a theory of justice? The Journal of Philosophy 103: 215-38
Piketty, T (2014) Capital in the twenty-first century. Belknap Press, Cambridge
Acemoglu D, Robinson JA (2015) The rise and decline of general laws of capitalism. Journal of Economic Perspectives 29:3–28
Zhang WB (2006) Growth with income and wealth distribution. Macmillan, London
Zhang WB (2020a) The general economic theory: An integrative approach. Springer International Publishing, Switzerland
Piketty, T (2020) Capital and ideology, translated by Goldhammer A. Belknap Press, Cambridge
Fukuyama F (1995) Trust: The social virtues and the creation of prosperity. The Free Press, New York
Kikuchi, M. (2018). Public trust in government, Japan. in Farazmand A (ed). Global encyclopedia of public administration, public policy, and governance. Springer, Berlin.
Tabellini, G. (2010). Culture and institutions: Economic development in the regions of Europe. *Journal of the European Economic Association* 8::677–716.
Tabellini G (2008) The scope of cooperation: Values and incentives. The Quarterly Journal of Economics 123: 905–950
Marini, A. (2016). Cultural beliefs, values and economics: A survey. MPRA 69747.
Cees van der E (2018) The essence of politics. Amsterdam University Press, Amsterdam
Nissani M (1992) Lives in the balance: The cold war and American politics, 1945-1991. Harvard University Press, Cambridge

Chapter 3
Bifurcations, Chaos, and Unpredictability

> Everything has beauty, but not everyone sees it.
> Confucius

Chaos can occur in a system that is controlled by a very simple deterministic equation. Modern mathematics reveals the complexity of order and chaos with simple behavioral mechanisms. There is no need for any exogenous force to experience chaos. Chaos can continue permanently even without any exogenous forces. For instance, China, which had been mainly isolated from any main civilizations, had repeatedly experienced chaos and poverty with the simple mechanism of the Malthusian trap. A natural propensity to many babies and limited land led to cyclical dynasties until Western civilization forced China to open its doors. The political-economic structures of China were controlled by similar not complicated political-economic mechanisms over a thousand years, but China was full of bifurcations and chaos. We now provide more examples of how chaos occurs in different socioeconomic systems with relatively simple mechanisms. Models with far more complicated interactions and multiple mechanisms are referred to in my general economic theory (Zhang 2020a).

3.1 Chaotic Inflations from Simple Economic Mechanisms

Inflation, which measures a general increase in the prices of goods and services in a given period in an economy, is a key macroeconomic variable that affects everyone in society. When the economy experiences high inflation, the amount of money one holds can buy fewer goods and services. When the inflation rate is high and income is not increased, those who depend on wage incomes may suffer much if nominal wages are invariant. Inflations can be caused by different forces such as the demand and supply of goods and services. The main factor is often owing to persistent excessive growth in the money supply by the government. When the economic system is "near" an unstable state, any changes in markets can lead to chaos. Inflation

W. Zhang, *Complexity Theory and Uncertainties*, Understanding Complex Systems,
https://doi.org/10.1007/978-3-031-42394-9_3

enhances the opportunity cost of holding money. Uncertainty over future inflation may discourage saving and investment. Inflation may also help the economy to reduce unemployment, encourage loans and investment instead of money hoarding, and allow the government in carrying out monetary policy.

There were many periods of hyperinflation, during which paper money can become so valueless that it is burned to heat rooms like once in Germany or China. A relatively well-recorded and analyzed period of high inflation is referred to as the Great Inflation (1965–1982) in America, as shown in Fig. 3.1. It lasted nearly two decades. In the initial stage, the inflation rate was about 1 percent and then started to rise, reaching more than 14 percent, and eventually declining to about 3 percent by the end of the period. Economists have mentioned many factors over the years for explaining the phenomenon. No convincing conclusion is made. Perhaps, no one could identify the cause for the phenomenon as chaos theory tells us that even small changes in some policy parameters can lead to great changes. During the period, there were many changes in policies and the global economic environment. Federal Reserve policies, for instance, were allowed to make excessive growth of the money supply. Keynesian theory, which is the most popular macroeconomic theory, holds that the government can sustain a healthy economic system by managing aggregate demand through the spending and taxation policies of the fiscal authority and the monetary policies of the central bank. Especially, during the Great Inflation, it was commonly believed that there was a stable relationship between unemployment and inflation (which is proved invalid in general situations). The theoretical relation implies that if the inflation rate is kept properly high, unemployment is certainly low. Faith is free but faith leads to wrong behavior if applied improperly.

In modern capitalist societies, governments play an active role through monetary policies and taxation, in addition to many other methods such as investment in education, infrastructure, military expenditures, and the provision of various social welfare. Central banks use monetary policy rules to control economies. These rules are often targeted at inflation, unemployment, and economic growth. A modern economy is characterized by the complexity of networking with greatly reduced constraints of time and space owing to the rapid globalization of digital technology. Governments know, perhaps, what they want but seldom achieve them through their monetary

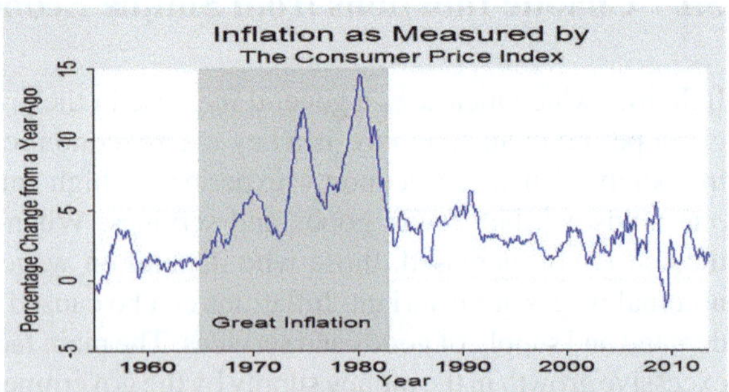

Fig. 3.1 The Great Inflation in the United States, 1965–1982. *Source* Michael Bryan (2013), https://www.federalreservehistory.org/essays/great-inflation

policies. Inflation is a phenomenon of interdependence between multiple cultural, psychological, political, and economic, as well as political forces.

There are many models built on simple economic mechanisms, which exhibit chaotic inflations (e.g., Zhang 2009. 2020b). I now illustrate a simple economic model which shows that a Keynesian economy with monetary policy exhibits chaotic behavior under certain conditions. I now introduce a model by Guevara and Escot (2021) to illustrate how three simple equations display chaotic inflation and unemployment. I simplify the model by directly specifying some parameter values. The model is composed of three equations. The first one describes the interdependence between the current inflation rate, the expected inflation rate, and the level of unemployment:

$$\pi_t = -1.14 + \frac{5.52}{u_t} + \frac{3.68}{u_t^2} + \pi_t^e,$$

where π_t and π_t^e are the current and expected inflation rates in period t, u_t is the level of unemployment. This is a version of the Phillips curve. The equation tells that the current inflation is positively proportional to the expected inflation and is nonlinearly negatively related to the current level of the expected inflation rate. It is empirically confirmed that a rise in the level of unemployment is associated with a fall in the inflation rate. The second equation is:

$$\pi_{t+1}^e \pi_t^e = 0.1\left(\pi_t - \pi_t^e\right).$$

It implies that the difference between the expected inflation rate for the next period and the expected inflation rate for the current period is positively proportional to the difference between the current inflation rate and the expected inflation rate for the current period. It is adaptive learning in the formation of expectations. If one does not make correct expectations in this period, one would form the expectation for the next period according to how one made inaccurate expectations. The third equation is:

$$u_{t+1} - u_t = -0.1(m - \pi_t),$$

in which m is the inflation policy determined by the central bank. The term $0.1(m - \pi_t)$ implies that an increase in the real amount of money increases the level of production, leading to a fall in the level of unemployment.

The economic system is composed of three equations. The system is unstable and exhibits chaotic behavior. Fig. 3.1 provides the bifurcation diagram of the unemployment rate with $m(11 < m < 17)$ as the bifurcation parameter. The figure and the rest of the figures in this section are from Guevara and Escot (2021). I consider that the system starts from the following state: $\pi(0) = 0.5$, $\pi^e(0) = 11$, and $u(0) = 0.8797$. It starts with an inflation rate equal to 0.5, an expected inflation rate of 11, and a level of unemployment of 0.8797.

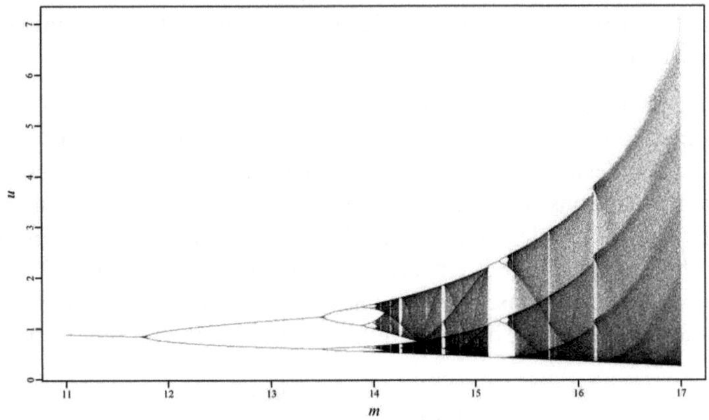

Fig. 3.2 Chaos with the inflation policy as the bifurcation parameter

The figure gives a typical route to chaos. As m increases, a period 2 limit cycle emerges. As the growth rate of money is further increased, a period 4 limit cycle appears. The cascade of bifurcations continues to increase till to chaotic dynamic equilibrium. Fig. 3.3 plots the behavior at $m = 16$.

One question is what one can do if the system is chaotic. Whether it is possible to escape from the chaos. Guevara and Escot apply the Taylor rule to see how the central bank can escape from the chaos. The well-known Taylor rule is a guide for driving policy instruments. It guards how central banks should act to achieve their macroeconomic objectives. The Taylor rule irefers to how central banks should determine nominal interests as economic conditions are changed. The rule was initially proposed by Taylor (1993) and Henderson and McKibbin (1993). It has helped central banks

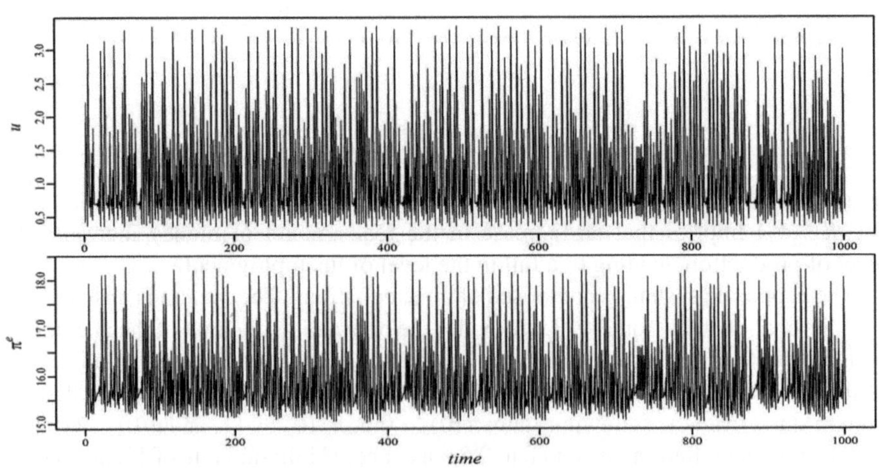

Fig. 3.3 Chaos in unemployment and expected inflation

in practice. Intuitively as in economic reasoning, one might consider the central bank to adjust the short-term nominal interest rate r according to the deviations from its targeted values of real interest rate r^*, inflation rate π^*, and equilibrium output level Y^*:

$$r - r^* = F\left(\pi - \pi^*, Y - Y^*\right)$$

where Y is the current level of output. If the inflation rate is higher than its target value, the central bank should raise its nominal interest rate for a fixed r^*; and if the current level of output is higher than its target value, the central bank should increase the nominal rate of interest. This relation was already applied before Taylor introduced the popular rule. Taylor considered that the nominal interest rate should be dependent on the deviations of actual inflation from its targeted level and actual GDP from its potential level. This idea is formulated as that the Federal Reserve should enhance interest rates when inflation is high or when employment is higher than full employment levels; on the other hand, the Federal Reserve should reduce interest rates when inflation and employment levels are low. Guevara and Escot apply the Taylor rule to control the chaos. They specify the following Taylor rule:

$$m_t = 16 - 0.07176911\left(\pi_t^e - 16\right) - 23.1343(u_t - 0.6960655).$$

As demonstrated in Fig. 3.4—which is the same as in Fig. 3.3 before the government applies the Taylor rule from $t = 400$—the government can achieve its long-term goals as shown in the trajectory in red which moves toward the targeted point $\pi^* = 15$ and $u^* = 0.696.u^* = 0.696$.

This example shows that chaos theory can also be applied to avoid unpredictable or undesirable behavior in the system. A typical example of the modern economy

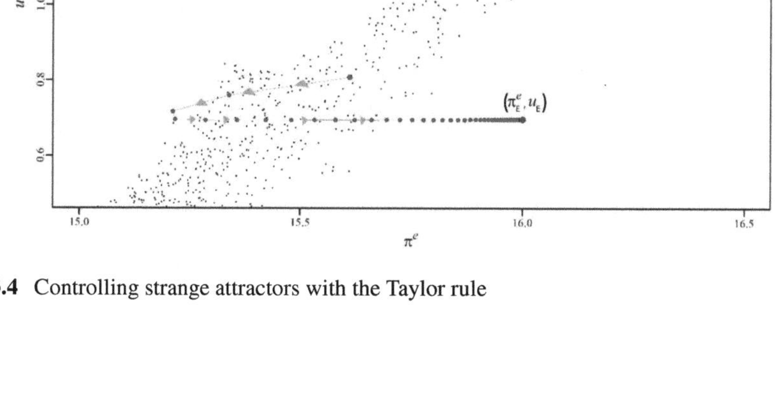

Fig. 3.4 Controlling strange attractors with the Taylor rule

is China's economic reform. Having been trapped in chaotic poverty, the Chinese government introduced freedom in markets for goods and services and in human relations in sex, marriages, and divorce with political dictatorship, economic planning, and state enterprises. In the last 45 years, China has achieved economic success with a relatively stable social environment, even though China is still under single-party control. Political stability and social structural stability were largely in order and the economic system has experienced chaotic but rapidly positive growth.

3.2 A Duopoly Game with Bounded Rationality and Behavioral Chaos

Various markets coexist in a modern economy. Students study perfect competition, imperfect competition, oligopoly, monopoly, and various games in the same course in microeconomics. Modern economic theory shows that economic chaos exists in economies with different mechanisms. We now show the possible existence of chaos in an oligopolistic market in which the trade is dominated by a small number of firms producing homogeneous products. Oligopoly markets are complex because firms must consider market demand and their own and rivals' strategies. A duopolistic model is proposed by Zhao et al. (2019) who study a dynamic multimarket Cournot model to analyze the behavior of oligopolistic producers, applying an incomplete information approach to generalize the traditional approach in which each producer applies the expectation to suppose that the opponents' output keeps the same level as previous period's and adopts an output strategy to maximize the expected profit. The system exhibits different movements. A similar extension to a duopoly Stackelberg game and identification of chaos is conducted by Yang et al. (2019). We now introduce a remanufacturing model with duopoly by Peng et al. (2019). The figures are from the same source.

Remanufacturing is a process in which used products are dissembled and some of their parts are utilized in the production of new products. There are many processes of remanufacturing in contemporary companies, such as Xerox, Mercedes-Benz, and Ford. Peng *et al.* model remanufacturing dynamics with two firms. There is interdependence between the firm, called firm 1, which produces original equipment products, and another firm, called firm 2, which collects and remanufactures the used product originally sold by firm 1. The competition between the two firms can be analyzed as a duopoly game. Firm 1 produces new products and firm 2 produces remanufactured products. Production decisions are made at discrete time $t = 0, 1, \ldots$. In period t, firm 1 manufactures and sells new products. These products can be returned and remanufactured by firm 2 in period $t+1$. Each consumer's willingness to—pay (WTP) for a remanufactured product is a fraction θ of their WTP for firm 1's product. Consumers' WTP is heterogeneous and uniformly distributed in the interval [0,1]. The linear inverse demand functions for the two firms are given as follows:

$$p_1(t) = 1 - q_1(t) - \theta q_2(t),$$

$$p_2(t) = \theta(1 - q_1(t) - q_2(t)),$$

where p_j and q_j are firm $j's$ price and quantity at time t, respectively. The total cost function of firm j is:

$$C_j = c_j q_j^2(t),\ j = 1, 2,$$

where are positive c_j positive parameter. The single period profit of firm j is:

$$\pi_j(t) = p_j(t)q_j(t) - C_j(t).$$

From the first-order conditions, we determine the optimal decisions of each firm. In practice, firms may not get complete information and do not have complete knowledge of the market. Firms make decisions with bounded rationality. It is considered that firms update output on the basis of the expected marginal profit of the current period. The dynamic adjustment is specified as follows:

$$q_j(t+1) = q_j(t) + v_j q_j(t)\frac{\partial \pi_j(t)}{\partial q_j(t)},\ j = 1, 2,$$

where v_j are positive constants. It can be shown that the system behavior is given by:

$$q_1(t+1) = q_1(t) + v_1 q_1(t)[1 - 2(1 + c_1)q_1(t) - \theta q_2(t)],$$

$$q_j(t+1) = min[f(t), q_1(t)],$$

where

$$f(t) = q_2(t) + v_2 q_2(t)[\theta - 2(1 + c_2)q_2(t) - \theta q_1(t)].$$

The above system describes a duopoly game played by boundedly rational players deciding on a process of dynamic investment in the market. The system exhibits complicated behavior with different parameters.

The system has five parameters, θ, c_j and v_j. Consider the case that the output traces of the two manufacturers are the same. It is sufficient to display firm 1's output for illustration of complex behavior. Fig. 3.1a shows the bifurcation diagram of q_1 with regards to the speed parameter v_1 when the other parameter values are specified as: $c_1 = 0.4$, $c_2 = 0.1$, and $\theta = 0.85$. When the speed is low, the system tends to be stable. As the speed is enhanced, the system becomes unstable, bifurcates, and the two-period solution appears. Fig. 3.5b shows the bifurcation diagram of q_1 with

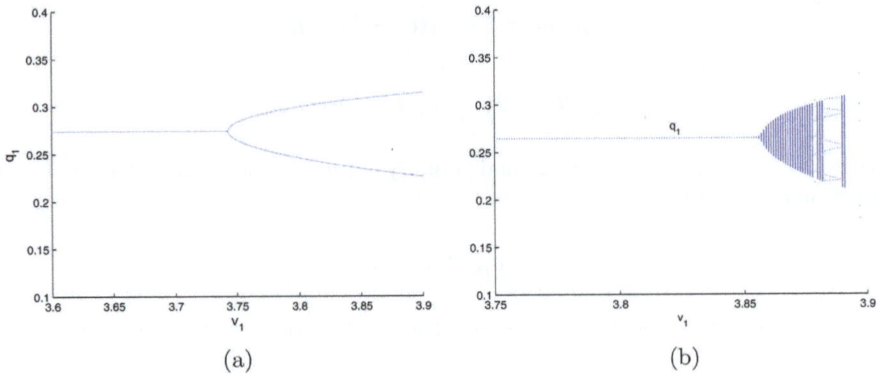

Fig. 3.5 The bifurcation diagram with v_1

regards to the speed parameter v_1 when the other parameter values are specified as: $c_1 = 0.4$, $c_2 = 0.17$, and $\theta = 0.98$. The authors identify period-doubling bifurcation and final chaotic behavior.

Figure 3.6 shows the bifurcation diagram of q_1 and q_2 with $v_2 = 4.4$ for all the cases. Fig. 3.6a provides the bifurcation diagram concerning v_1 when the other parameter values are specified as: $c_1 = 0.4$, $c_2 = 0.1$, and $\theta = 0.6$. It is observed that a period-doubling bifurcation occurs at $v_1 = 1.2121$. Fig. 3.6b is the bifurcation diagram concerning θ when the other parameter values are specified as: $c_1 = 0.4$, $c_2 = 0.2$, and $v_1 = 1$. As θ increases, the equilibrium becomes unstable and complex dynamic behavior ours. Fig. 3.6c is the bifurcation diagram concerning c_1 when the other parameter values are specified as: $\theta = 0.6$, $c_2 = 0.2$, and $v_1 = 1$. Two-period bifurcation occurs. Fig. 3.6d is the bifurcation diagram with respect to c_2 when the other parameter values are specified as: $\theta = 0.6$, $c_1 = 0.4$, and $v_1 = 1$.

3.3 Housing Market Crashes and Catastrophe Theory

Housing market crashes occur in different countries. For instance, in the United States, there were four major housing crashes. They happened respectively, in the panic of 1837, the 1873 stock market crises, the 1929 Wall Street crash, the Great Depression of the 1930s to 1940s, and the 2008 housing bubble. Fig. 3.7 compares the housing market crashes in the 1990s in Japan and the 2000s in America. There are many similarities between the two crashes from many perspectives. There are many speculations about crashes in China's housing markets. Crashes are not predictable in detail, but each crash occurs in certain structures under certain conditions. Structural instabilities which are determined by interdependence between many variables are basic features of bubbles.

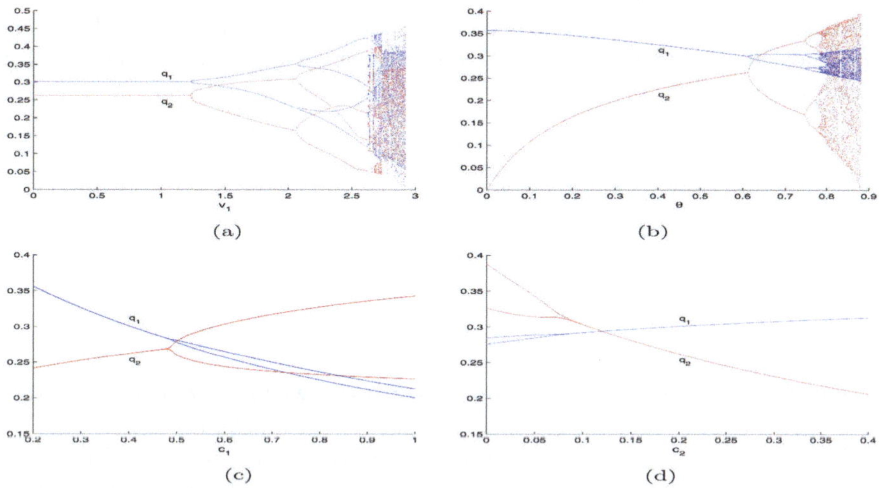

Fig. 3.6 Bifurcations and chaos

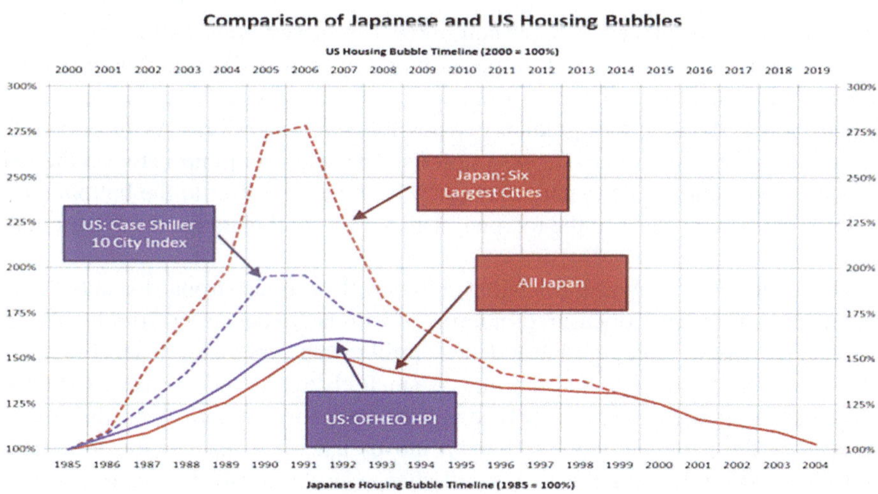

Fig. 3.7 Housing market crashes in Japan and America. *Source* https://seattlebubble.com/blog/2008/11/03/comparing-the-us-and-japanese-housing-bubbles/

Chaos theory provides some insights into the mechanisms of bubbles and their predictability in consequences. I now introduce the study by Dikes and Wang (2016) who explain some observed abrupt changes in housing markets by the stochastic catastrophe model. Catastrophe theory is proposed by Thom (1972). The theory is extensively applied to different fields (e.g., Zeeman 1977; Zhang 1991, 1997, 2003a; Guastello 2001; Wilson 2011). The theory is initially for deterministic systems. It is soon generalized with random factors. For a deterministic system, consider the cusp catastrophe, which is most commonly applied in behavioral science. Dikes and Wang consider a potential function V_t with a single variable z_t and two parameters α and β :

$$V_t = \frac{1}{4}z_t^4 - \frac{1}{2}\beta z_t^3 - \alpha z_t.$$

Equilibrium points are determined by:

$$\frac{\partial V}{\partial z} = z^3 - \beta z - \alpha = 0.$$

With Cardan's determinant $\delta \equiv 27\alpha^2 - 4\beta^3$, as shown in Fig. 8 (Dikes and Wang 2016), if $\delta < 0$, there are three solutions; while if $\delta > 0$, there is one solution. The figure displays the cusp equilibrium surface in a three-dimensional space. The folded surface with a cusp fold shows the equilibrium surface of the equation. The floor is a two-dimensional control plane with two parameters. The height represents the value of the equilibrium state for varied combinations of the control parameter values. The center of the graph displays three sheets: the middle one shows the unstable equilibrium and the other two stable equilibria. The curve formed by the edges of the fold cusp projected onto the control space forms a cusp-shaped region. The cusp is the bifurcation set, determined by $\delta = 0$. When $\delta > 0$, there is a stable equilibrium point. When one $\delta < 0$, the surface predicts two possible stable equilibria. As the parameter values vary, the system may show hysteresis. The state can jump between the two possible state values. The system may jump from the top sheet to the bottom sheet or from the bottom sheet to the top one, with varied combinations of the parameter valves.

Dikes and Wang analyze the housing market with cusp catastrophe by considering the housing price level (measured in terms of its relative deviation from the fundamental price) as the state variable and the control parameters as, for instance, interest rate, government policies, and the like. A small change in the control parameters can create a great change. For instance, if the state follows path A on the control surface, as the parameters are shifted, it moves left till it reaches the fold curve and jumps to the bottom sheet. A small change in the parameters causes a great change in the state variable. Path B outside of the cusp bifurcation does not experience any sudden change when it moves from the top to the bottom. Parameters or environment determines when the system experiences sudden changes or not. When applying this model to housing markets and considering housing price as the state variable, it can be seen why crashes may occur if people had little idea or information about the values of control parameters. The model also implies that if an economy has sufficient knowledge and information, it can avoid "negative" chaos and sudden changes. No country in the modern world has proper information about why and how households, companies, and governments make their decisions. All open societies nowadays unpredictably experience different types of chaos and sudden changes.

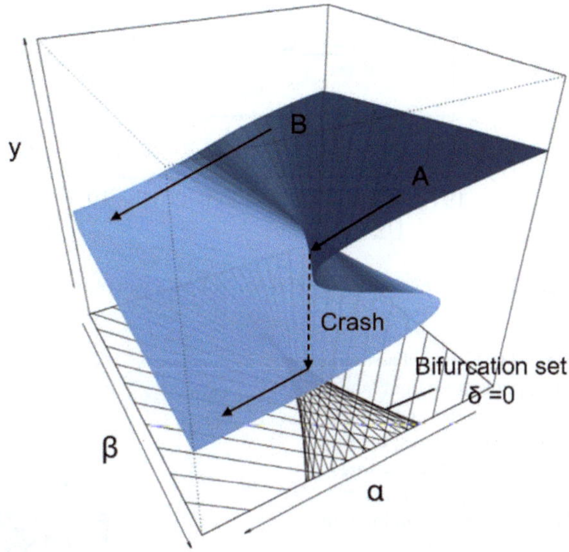

Fig. 3.8 The cusp catastrophe

Dikes and Wang use data from different countries and test their model. They estimated behavior points of different countries with regards to control parameters α and β, as shown in Fig. 3.9. The empirical study confirms that the cusp catastrophe model can effectively explain what has been observed in the housing markets of different economies.

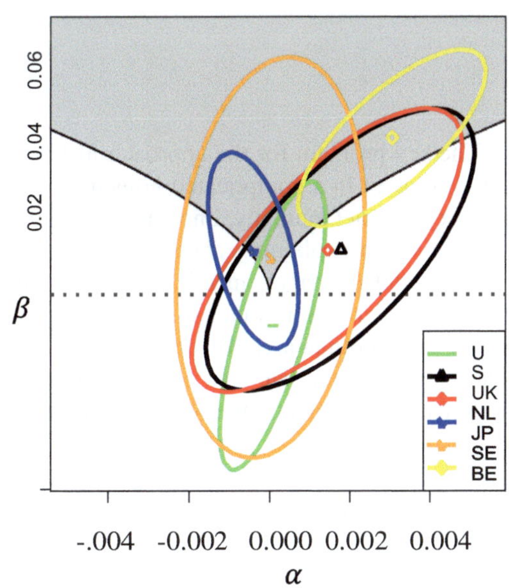

Fig. 3.9 The control parameters values for Japan and America

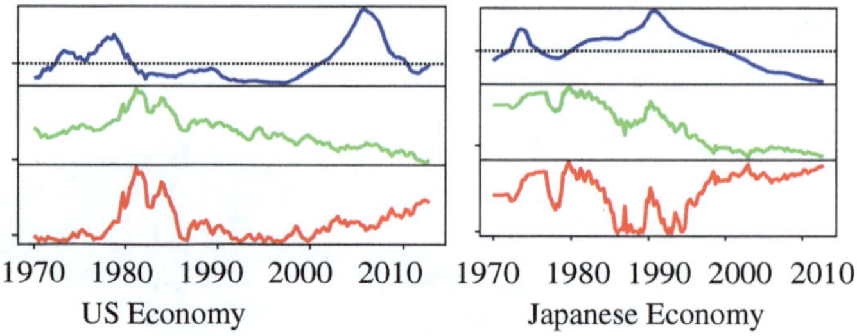

Fig. 3.10 The state variables and control parameters for Japan and America

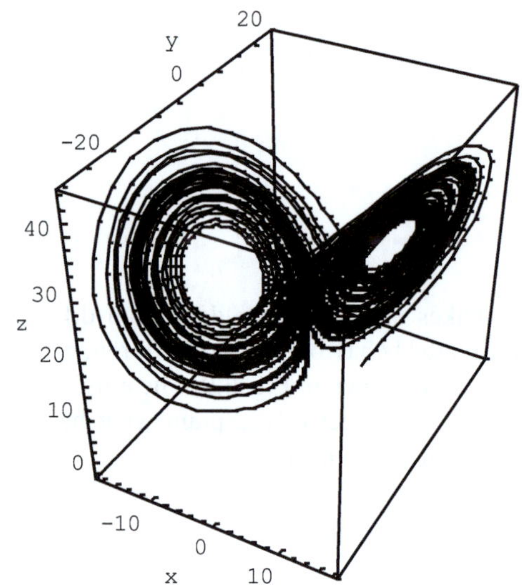

Fig. 3.11 Urban chaos with GDP, population, and land rent. *Note* vertex data reflect the relative changes; the blue line stands for the housing price m, the green line for the interest rate, and the rate for the value of $\delta^{\frac{1}{3}}$

The key problem for the approach just described is that the system has a single state variable. In a complicated housing market, there are multiple state variables, and they interact mutually. I now use my 1991 model to illustrate another economic mechanism in explaining chaos in housing markets (Zhang 1991, Chap. 6). The urban model describes an urban system within a metropolitan area. The system under consideration is small so its dynamics will have almost no significant impact on the metropolitan area. Businesses and residents are free to choose their location sites either in the urban area or other areas. Locational characteristics of the urban system are described by the following three variables:

$$X(t) = \text{the output of the urban system at time } t;$$
$$Y(t) = \text{the number of residents}; Z(t) = \text{the land rent.}$$

I Propose the Following Urban Dynamics:

$$\dot{X} = a_1 Y - a_2 X,$$

$$\dot{Y} = c_1 X - c_2 Y - c_3 XZ,$$

$$\dot{Z} = -d_2 Z + d_1 XY,$$

where a_i, c_i, and d_i are positive parameters. The economic interpretations of the model are referred to by Zhang (1991, Chap. 6). The first equation implies that growth in the city's GDP is positively related to its population size and negatively related to its current level. The second equation population growth is positively related to its output and negatively relative to its current population. The third equation tells that change in the urban land rent is negatively related to its current level but positively dependent on the population and the output level. Under linear transformations, the system is identical to the celebrated system of equations that Lorenz (1963). The Lorenz equations are a quadratic system of autonomous differential equations in three dimensions modeling a three-mode approximation to the motion of a layer of fluid heated from below. A thorough treatment of the Lorenz equations is given by Sparrow (1982). As demonstrated in Fig. 3.12, the urban system exhibits chaotic behavior (with rescaled variables with proportional transformations). There are two sheets in which trajectories spiral outward. When the distance from the center of such a spiral becomes larger than some threshold, the motion is ejected from the spiral and is attracted by the other spiral, where it again begins to spiral out, and the process is repeated. The motion is not regular. The number of turns that a trajectory spends in one spiral before it jumps to the other is not specified. It may wind around one spiral twice, and then three times around the other, then ten times around the first, and so on.

The Lorenz attractor is dubbed a strange attractor because there are no asymptotically stable equilibria or period orbits in a global attractor that is a compact, connected invariant set. The geometry of the attractor is exceedingly complicated. We may also visualize projects in $xz-$ plane as displayed in Fig. 3.12. The graphs appear to cross over themselves repeatedly, but this cannot be true for the actual trajectories in three-dimensional space because of the unique theorem. The apparent crossings are due wholly to the two-dimensional character of the figure.

The urban system exhibits the following properties: (1) the temporary path of the three urban variables is time dependent but is not periodic (or "regular"); (2) the motion does not appear to show a transient phenomenon since, regardless of how long the numerical integration is continued, the trajectory is going to continue to wind around and around without settling down to either periodic or stationary behavior; (3) the topology of the figure is not dependent on the choice of initial conditions or integrating route; and (4) it is impossible to predict the details of how the trajectory will develop over any period other than a very short time interval. Considering the above urban interpretation of the Lorenz equations, I show that even if the government is well-informed and is composed of well-educated officials and

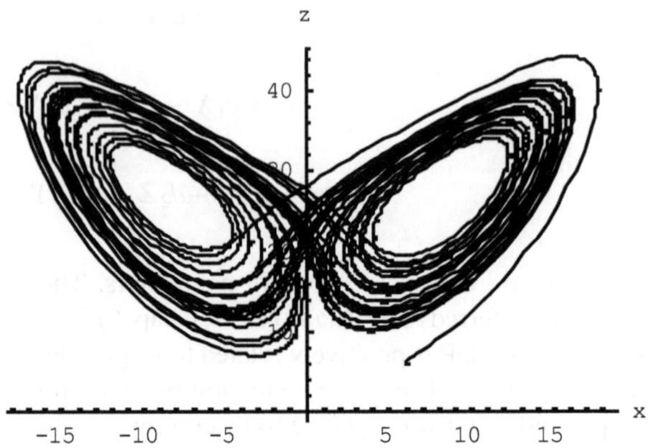

Fig. 3.12 Projections of a trajectory of the Lorenz equations

experts, the government can't predict the impact of its actions, such as tax policy, land policy, and infrastructure policy.

3.4 Delays in Actions and Business Cycles

There are different interpretations of time-dependent phenomena. An early approach to modeling is owing to Kalecki (1935) and Kaldor (1940, see also Zhang 1990). Kalecki considers that there is a delay between an investment decision and the delivery of the investment. When modeling economic growth, this idea is reflected in a differential equation with delay. The model can generate periodic oscillations. Similarly, in any field related to human decisions and actions with multiple agents, timing and delays, or the opposite (with regards to the best timing) are crucial for revealing the complexity of the human network. Kaldor explains the existence of business cycles with nonlinear relations in investment and saving in economic growth. Krawiec and Szydlowski (1999) combine the ideas of Kalecki and Kaldor and develop the Kaldor–Krawiec business cycle model. Yu and Peng (2016) consider the business cycle model with both discrete and distributed delays. I now introduce their simulation results. The figures in the rest of the section are from the same source. The model consists of two differential equations with a discrete delay and a distributed delay in the form of:

$$\dot{Y}(t) = \alpha(I(Y(t), K(t)) - S(Y(t), K(t))),$$

$$\dot{K}(t) = I\left(Y(t-\tau), \int_{-\infty}^{t} F(t-s)K(s)ds\right) - \delta K(t),$$

in which Y is the gross production, K is capital stock, I is an investment, national income, S is savings, δ is depreciation rate of capital, α is a positive parameter. Here, $\tau > 0$ is the investment delay or expected time of capital stock. The function $F(s)$ is a non-negative continuous delay kernel defined and integrable on $[0, +\infty)$. It measures the influence of past states on current dynamics. The function is specified as:

$$F(s) = \alpha_1 e^{-\alpha_1 s}, \alpha_1 > 0.$$

In the numerical simulation, the functional forms and parameter values are specified as follows:

$$L(t) \equiv \int_{-\infty}^{t} F(t-s)K(s)ds, I = \frac{1 + exp(Y)}{exp(Y)} - 0,2K,$$

$$\alpha = 3, \alpha_1 = 0.6, \gamma = 0.2, \delta = 0.1.$$

It is straightforward to check that the dynamic system can be rewritten:

$$\dot{Y}(t) = \frac{1 + exp(Y)}{exp(Y)} - 0,6K - 0.6K(t),$$

$$\dot{K}(t) = \frac{1 + exp(Y(t-\tau))}{exp(Y(t-\tau))} - 0.2L(t) - 0.1K(t),$$

$$\dot{L}(t) = 0.6K(t) - 0.6L(t).$$

The system has a unique positive equilibrium (1.31346, 2.62699, 2.62699). The delay is chosen as a bifurcation point. When $\tau_0 = 4.4019$, Hopf bifurcation occurs. Fig. 3.13 displays business cycles when $\tau_1 = 4.5 > \tau_0$. The positive equilibrium point loses stability when the delay is lengthened. Business cycles appear as decision-making processes are changed.

3.5 Population Dynamics and Malthusian Economy

Population growth is essential for any society's sustainability. Some modern industrialized economies are now faced with challenges caused by extremely low birth rates. Societies fail to reproduce in terms of population. On the other hand, Malthus is well-known for his dismayed conclusions about economic development and population growth. Overpopulation is the main concern of Malthusian population theory. Malthus is not the first to examine demographic problems, but he is the first who succeeded in developing a theory of population growth. By exploring the interdependence of population growth and food supply, Malthus argues for the subsistence

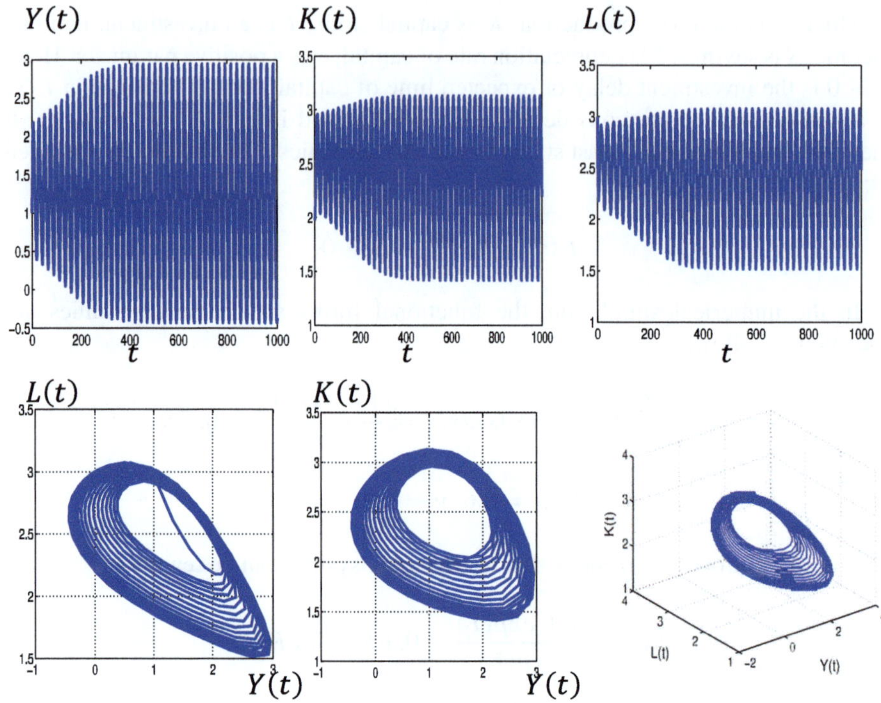

Fig. 3.13 Business cycles as the delay is lengthened

theory of wages. The idea has important influences on Ricardo, Marx, and Keynes. He explains poverty in terms of a simple interdependence between the population and the means of subsistence. The theory still provides valid insights into the mechanisms of global poverty. One well-known—not so human—example to control overpopulation in a poor economy through compulsory control is China's one-child policy. The basic content of the policy is that each Han Chinese woman—if not remarried and has a healthy child or twin—is allowed to bear only one baby in a lifetime. If the couple has two, the mother and the father are punished in varied ways according to their social positions. China is composed of 56 races. Han is dominant. The one-child policy was applied only to Han people. The other races—such as Korean Chinese and Islam Chinese—were permitted to have two children. The policy was abolished some years ago because the economic reform resulted in the typical phenomenon of modern industrialized economies with gender freedom that per woman's lifetime birth number is below the replacement rate of 2.2. China's simple rule has unbelievable consequences for China and the world (especially, the global environment and resource utilization and distribution). The number of people that the policy reduced is almost equal to the population of the United States (if all the other conditions are the same). The Global population is much reduced. If without the policy, one might wonder whether China could have succeeded in the economic reform or whether China might have suffered great human calamities.

I now illustrate the essence of the Malthusian theory. The remainder of this section is based on Zhang (2005, Sect. 1.2), In the simplest form of the Malthusian growth model, the population grows at a constant rate. There is no checking force on the birth rate. This implies that as far as people survive, they will have birthrates independent of any other social and economic forces. The population would grow at a natural rate. That is, $\dot{N}(t) = aN(t)$, where $N(t)$ is the population at time t and a is a positive parameter. This kind of population growth may be valid for a short time, but it clearly cannot go on forever. There are limitations of natural resources, for instance, that will not allow the population to grow without limit. The well-known logistic growth model is often applied to explain population growth. The logistic model is formed as:

$$\dot{N} = aN - abN^2 = aN(1 - bN).$$

Except for the natural growth tendency of reproduction, the population growth is subject to checking effects of resources, given by abN^2. Nevertheless, from a long-term perspective, resources for human societies are not given but are changeable owing to, for instance, technological changes. National production affects capacities to support the population. To analyze how output affects population growth, Haavelmo (1954) suggests the following extension of Malthus' system:

$$\dot{N} = aN\left(1 - \frac{bN}{Y}\right), a, b > 0$$

in which Y is the real output which is a function of technology A and the labor force N:

$$Y = AN^{\beta}, A > 0, 0 < \beta < 1$$

Substituting the production function into the population growth equation yields

$$\dot{N} = N\left(a - a_1 N^{1-\beta}\right),$$

where $a_1 \equiv ab/A$. The growth law is a generalization of the familiar logistic form widely used in biological population and economic analysis. As shown in Zhang (1991), the discrete version of the above equation creates chaotic behavior, as illustrated in Chap. 1. To simulate the model I specify the parameters:

$$a = 0.05, b = 0.4, \beta = 0.6, A = 1.4.$$

From the initial condition $N(0) = 8$, I run the model for 100 years. Fig. 3.14 depicts the motion of $N(t)$, $Y(t)$, and the income per capita, $y(t)(\equiv Y(t)/N(t))$.

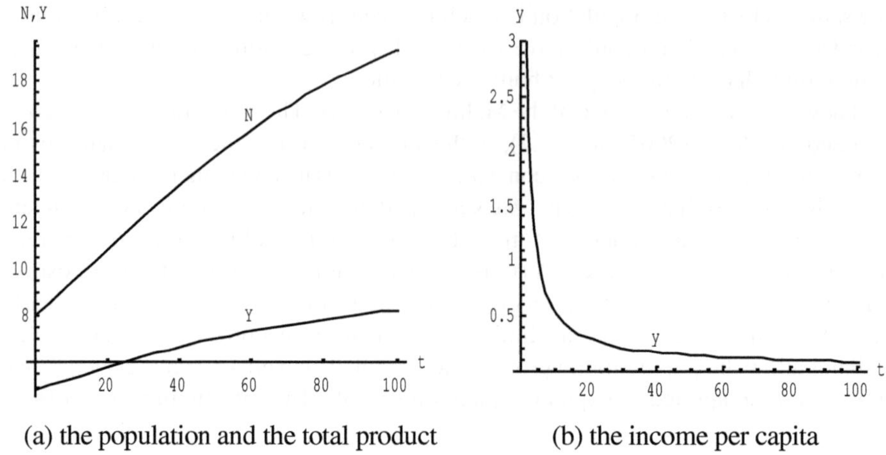

Fig. 3.14 The dynamics of a Malthusian economy

As shown in Fig. 3.14, the national product grows over time, but the income per capita first suddenly but afterward steadily declines. As observed in modern economic history, population growth is associated with improvement in living standards. An obvious way to avoid the dismaying implications of the Malthusian theory is to consider technological change. For illustration, I specify a pattern of exogenous technological progress as follows:

$$A(t) = 1.4 \exp(0.02t).$$

As time passes, the total output will be enhanced with the same level of labor force owing to technological change. The technology parameter grows $A(t)$ annually by 2 percent. Fig. 3.14 shows the motion of the system with the technological change. Comparing the two figures, I observe that with the technological change, the economy dramatically changes over 100 years. The economy is enlarged and the income per capita grows as the population grows after an initial decline.

It should be noted that Malthus mentions uncertainties in economic evolution with endogenous population. He believes in nonlinear structural relations and their complicated multiconnected nature. Modern nonlinear sciences provide a much more valid and testable way to population problems.

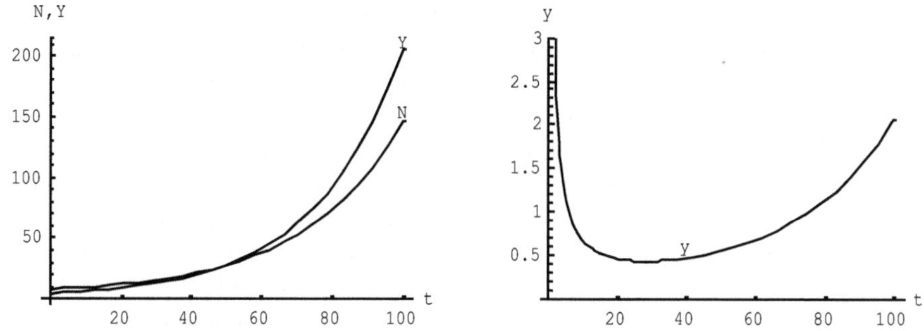

(a) the population and the total product (b) the income per capita

Fig. 3.15 The dynamics of the Malthusian economy with technological change

3.6 Chaos and Optimal Growth with Human Capital

Economic dynamics is to study changes in economic systems and the forces generating them. It is concerned with how, why, and when changes take place in living conditions. An economic theory should explain changes in production, consumption, resource allocation, trading flows, prices, and many other variables. Although classical economists are concerned with economic dynamics in terms of changes in population, capital, knowledge, division of labor, economic structures, and institutions, they do not construct formal (mathematical) economic systems. Some of the most important empirical predictions of, for example, Malthus' and Ricardo's theories have not been realized. The share of landowners does not seem to increase, and the population does not grow faster than output. The importance of agriculture relative to manufacturing declines markedly. The primary object of modern economic growth theory is to explain the movements in the output, employment, and capital stock of a growing economy and the interrelations among these variables. It explains the movements in the distribution of income among the factors of production. Modern growth theory provides a conceptual framework within which much more meaningful empirical research can take place. The economies which it attempts to describe are essentially advanced and industrialized. In such economies, capital, technology, and labor are the three inputs upon which attention is focused. Land, which is an important input in the classical growth theory, is usually ignored.

Modern growth theory is often considered from Ramsey's article published in 1928 (Ramsey 1928). The works by Harrod (1939, 1948) and Domar (1946) make the long-run growth problems more rigorous and more statistically testable than the classical works. However, these models are crucially dependent upon the assumptions that the capital/output and capital/labor ratios are constant. This drawback is remedied by Solow (1956) and Swan (1956), who introduced production functions into growth theory. It is commonly held that neoclassical growth theory is initiated by the Solow–Swan model (often referred to as the Solow model). The model assumes constant returns to scale (diminishing returns to each input) and positive and smooth elasticity of substitution between the inputs. The model is attractive and popular because it has a simple economic structure and a unique stable steady state, with clear and positive implications on social welfare and market mechanisms. It fixes the saving rate, the growth rate of the population, and technology with a homogenous population. the position of the production function. The model predicts that in the absence of continuing improvements in technology, per capita growth must eventually cease. Another dynamic property of the model is conditional convergence, stating that the lower the starting level of real per capita GDP in comparison to the steady position, the faster the growth rate. The saving rate is significant in determining the levels of other economic variables. By the early 1970s, neoclassical growth theory was almost neglected as an active research field. For about one and half decades, macroeconomic research had been mainly concerned with short-term fluctuations and rational expectations. Since the mid-1980s, growth theory has experienced a new boom, emphasizing endogenous technological change Endogenous technological progress is considered an essential determinant of economic development.

It is difficult to study economic issues related to creativity and knowledge utilization as well as capital accumulation within a compact framework. Although recent models of endogenous growth are claimed to be related to neoclassical growth theory, most of these works neglect physical capital accumulation. It does not seem that the new approach to growth theory is an improvement of the neoclassical growth theory. These new models may not be more valid in explaining economic dynamics than the neoclassical growth models developed in the 1950–60s. The omission of capital accumulation is not due to the insignificance of capital accumulation in explaining modern economic growth but mainly because mainstream economists fail to construct an analytical framework by which issues related to economic growth and structural change with endogenous capital and knowledge can be analyzed consistently.

I now introduce a typical model to demonstrate the existence of chaos with endogenous physical capital and human capital subject to rationality, determinism, and optimization. I describe the Lucas model (Lucas 1988) reformed by Benhabib and Perli (1994) and then show numerical illustrations. Formally, the Lucas model is as follows:

$$\max_{c(t),u(t)} \int_0^\infty \frac{c^{1-\sigma}-1}{1-\sigma} e^{-\rho t} dt$$

Subject to:

$$\dot{k} = Ak^{\beta}(uh)^{1-\beta}h^{\gamma} - c,$$

$$\dot{h} = \delta h(1-u),$$

where c stands for consumption, k is physical capital, h is human capital, u is the work hours, and $1-u$ is the hours devoted to accumulating human capital. The first differential equation implies that the change in the capital is the output minus consumption. Since the physical description is neglected, what the economy produces is either invested or consumed. The second differential equation implies that the change rate of human capital is linearly proportional to the number of researchers. We have γ as a positive externality parameter, and A, β, σ, δ, and ρ are parameters. Benhabib and Perli introduce two variables:

$$x \equiv kh^{\frac{(1-\beta-\gamma)}{(\beta-1)}}, q \equiv \frac{c}{k}.$$

They proved that the motion of the Lucas model is rewritten in the following form of three nonlinear differential equations:

$$\dot{x} = Ax^{\beta}u^{1-\beta} + \frac{\delta(1-\beta-\gamma)}{\beta-1}(1-u)x - qx,$$

$$\dot{u} = Ax^{\beta}u^{1-\beta} + \frac{\delta(\beta-\gamma)}{\beta}u^2 + \frac{\delta(1-\beta-\gamma)}{\beta-1}u - qu,$$

$$\dot{q} = q^2 + A\frac{(\beta-\sigma)}{\sigma}x^{\beta-1}u^{1-\beta}q - \frac{\rho}{\sigma}q.$$

Bella et al. (2017) examine some nonlinear phenomena which are not revealed in the previous studies. Bifurcations occur on the parametric hypersurface as shown in Fig. 3.16 under some additional conditions. The figures in this section are from Bella et al. The rest of this section fixes:

$$A = 0.2, \delta = 0.05, \rho = 0.02297.$$

The parameters γ, β, and σ are taken on wide ranges of values.

Fig. 3.16 The parametric hypersurface for bifurcations

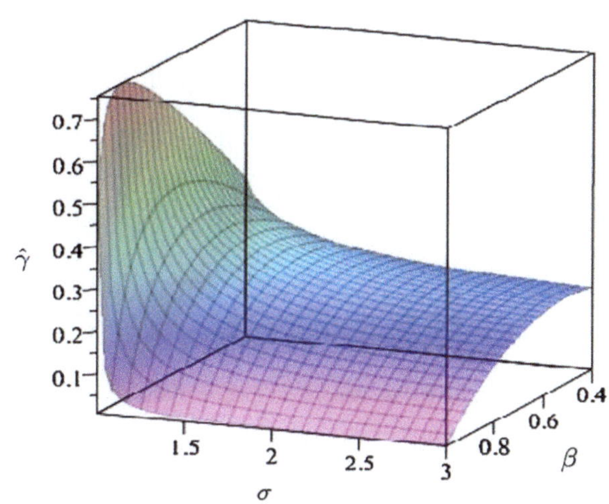

Let variable g denote the growth rate of the economy. Figure 17 shows irregular time-dependent behavior of the economy with the following parameter values.

$$\beta = 0.5, \gamma = \frac{1}{3}, \sigma = 2.5.$$

Bella, *et al.* also vary γ near the value in Fig. 3.17. Figure 3.18 shows how the system oscillates as γ is perturbed (red for $\gamma = 0.32$; green for $\gamma = 0.4$; blue for $\gamma = 1/3$).

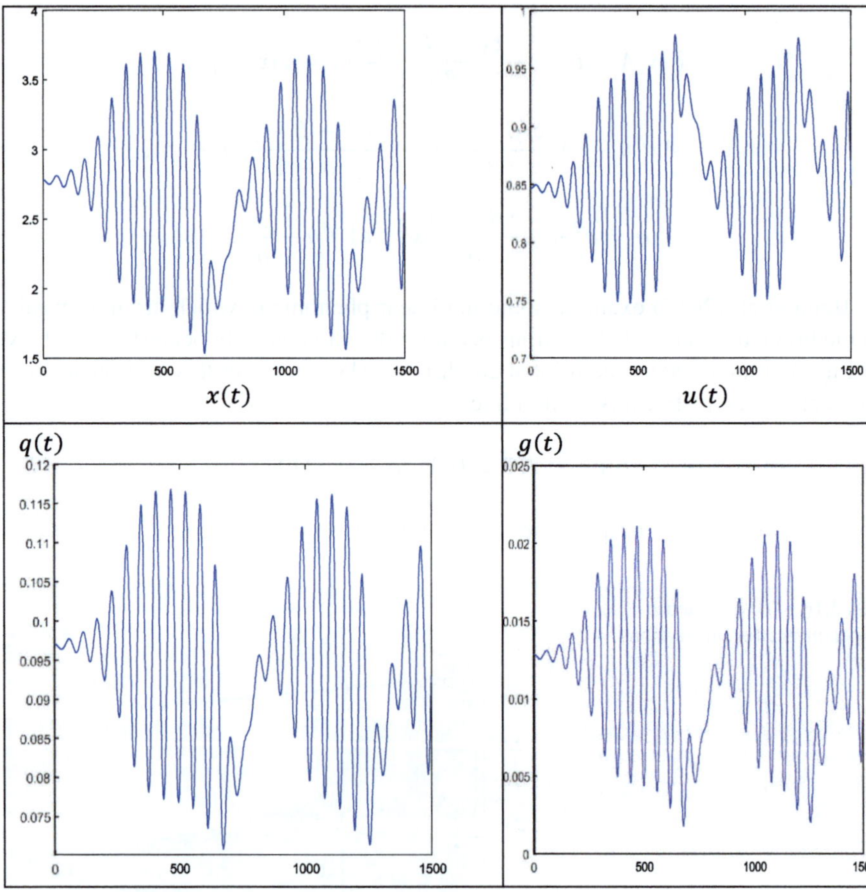

Fig. 3.17 Irregular time-dependent motion of the economy

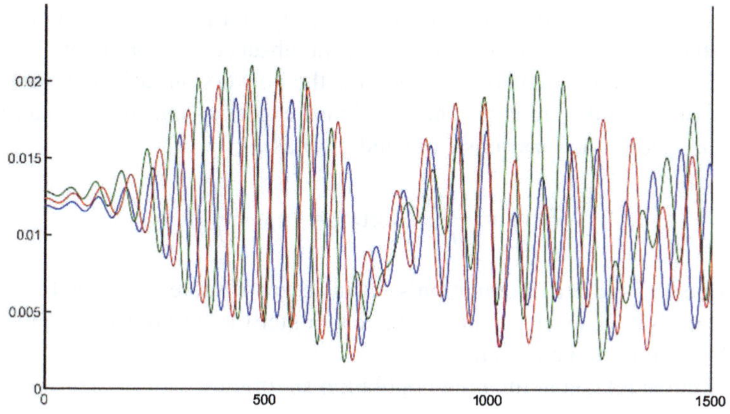

Fig. 3.18 Irregular growth rates with changes in γ

3.7 Prey–Predator Relations and Chaos

Relations between human races, animals, and natural environments are complicated. Races coexist in conflicts, cooperation, or actual separation from each other. For instance. Karl Marx characterizes—biasedly or perhaps wrongly—national history as a history of class struggle. How affairs are related to each other is basic to understanding the evolution of animals as well as humans. Adam Smith builds his economic world with the assumption that humans pursue their interests and that free markets lead to optimal social consequences. His theory does not deal with the coexistence of multiple races in population dynamics. In the near future, the White population may have little influence on global scales because of the decline of the population. In mathematical biology, a prey–predator model—often referred to as the Lotka-Volterra—plays an important role in revealing the complexity of interdependence between the two populations. Prey-predator models are a key framework in mathematical biology and are applied in various fields of social sciences. In the contemporary world, human efforts are required to sustain the coexistence of animals. Fishing is a typical example. I now introduce a prey–predator model, which is influenced by the literature on the economics of renewable resource management and harvesting (e.g., Clark 1976, 1985). The model and figures presented below are from Gupta et al. (2014). The model introduces the harvesting function proposed by Agnew (1979) to a prey-predator model.

The prey–predator model with harvesting by Gupta et al. (2014) is as follows:

$$\dot{x}(t) = xg(x) - y\overline{p}(x) - h(x, E),$$

$$\dot{y}(t) = y(-d + \theta\overline{p}(x)), x(0), y(0) > 0,$$

where $x(t)$ and $y(t)$ stand for the densities of prey and predator, the function $g(x)$ measures the per capita net prey growth in the absence of a predator, $\overline{p}(x)$ is the functional response of predator, θ stands for the conversion rate, and d means the natural death rate of the predator. The function E expresses the effort to harvest prey. Bella, *et al.* specify the functions $\overline{p}(x)$ and $g(x)$ as:

$$\overline{p}(x) = \frac{ax}{m+x}, g(x) = \mathrm{r}\left(1 - \frac{x}{k}\right),$$

where r refers to the prey population's intrinsic growth rate, k stands for the environmental carrying capacity, a means the encounter rate of predator to prey, and m is the half of saturation constant.

Harvesting is related to the prey population as follows:

$$h(x, E) = \frac{qEx}{sC_0^{-1} + E + bsx}$$

in which $E = sV$, s being a proportionality constant and V number of identical vessels per unit area for harvesting, C_0 is the degree of competition between the vessels, q is the catchability coefficient, and s is the proportionality constant for handling time. The effort is situation dependent:

$$\dot{E} = n\left(\frac{(p-T)h}{E} - c\right)E,$$

in which n is an adjustment parameter, p profit, T is the tax rate, and c is the cost per unit biomass.

Gupta et al. investigate the stability of various equilibrium points, Hopf bifurcation around axial equilibrium points and interior equilibrium points. In numerical examples, they specify the parameter values:

$$a = 0.25, r = 0.25, k = 100, m = 10, d = 0.05, p = 3, \theta = 0.3,$$

$$q = 0.3, C_0 = 1, b = 1, s = 1, c = 0.15, n = 1.$$

They choose the tax rate as the bifurcation parameter. Figure 3.19 shows that there are business cycles when $T = 2$. The figures in this section are from Gupta et al. (2014). In this section, the red curves represent the prey population, the green curves stand for the predator population, and the black curves stand for the effort for harvesting. If the tax rate is further increased, new bifurcations occur.

Figure 3.21 displays how a small difference in initial states results in long-term differences in the behavior of the system and gives the bifurcation diagram for $2.25 \le T \le 2.06$. In Fig. 3.21a, the green and red curves display the prey population with two different initial states at $T = 2.06$. Fig. 3.22b provides the bifurcation diagram, using the tax rate as the bifurcation parameter.

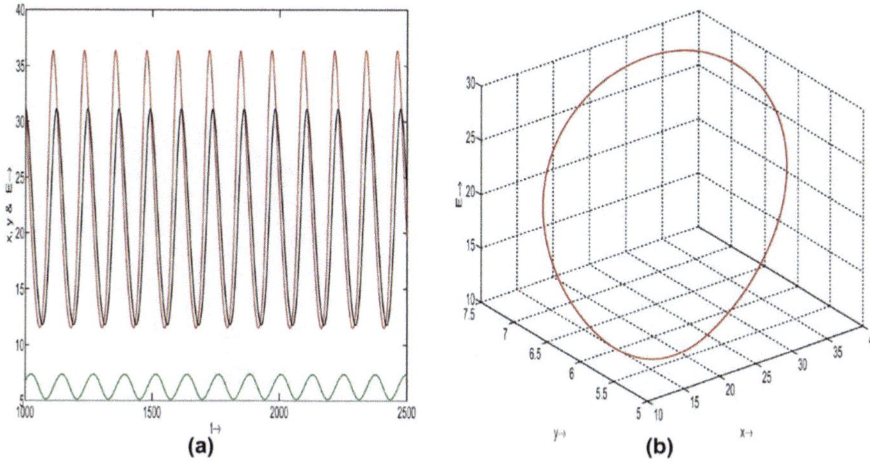

Fig. 3.19 The periodic populations and efforts when $T = 2$

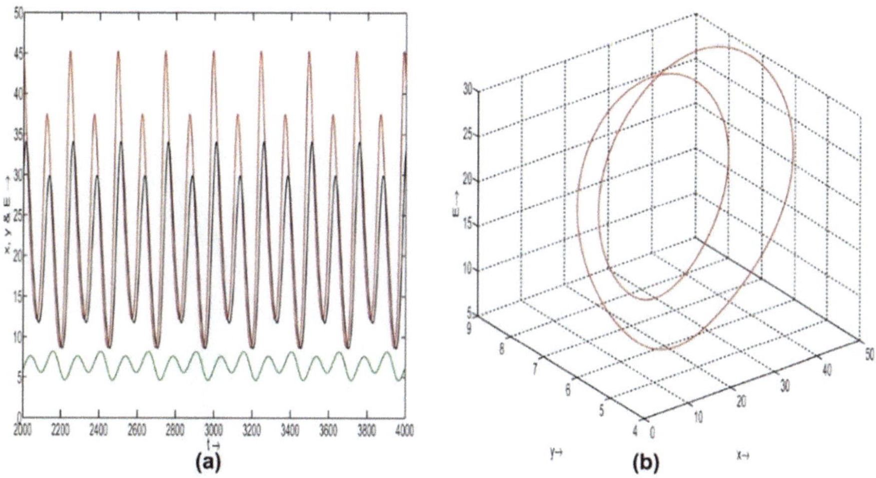

Fig. 3.20 The period 2 solution appears when $T = 2.03$

3.8 Business and Environmental Cycles

Mankind has never had such a capacity to destroy its living environment as today. Environmental changes threaten not only individual regions but even the Earth itself. The possible danger is associated with industrialization, increased consumption, and rapid population growth. Data from the World Bank displays that both global GDP per capita and global trends for air pollution (as measured by levels of PM2.5) follow a similar pattern of growth, as plotted in Fig. 3.23 (Dang et al. 2020).

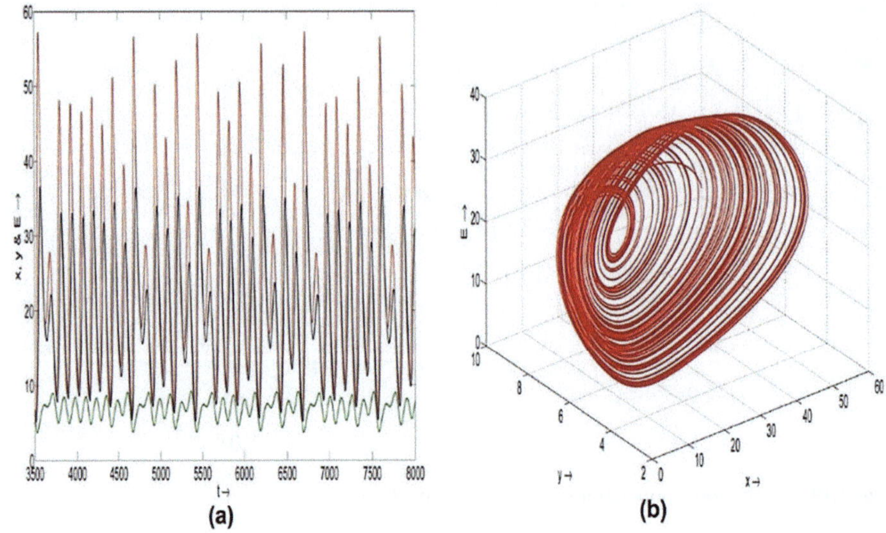

Fig. 3.21 Economic chaos when $T = 2.06$

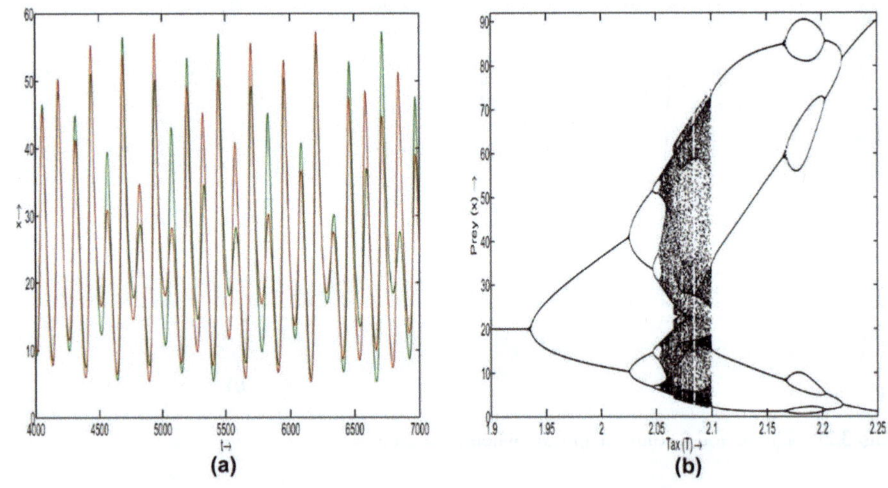

Fig. 3.22 Behavior with differential initial states and bifurcation diagram

Countries show great differences as illustrated in Fig. 3.24 (Dang et al. 2020). During the same period, for instance, Norway's environment was improved in association with economic growth, while China's economic growth was accompanied by a faster deterioration rate. There are no linear relations between environmental change and economic growth. In the case of China, air pollution has been improved from 2010 onward. China took strong measures, such as austere clean air regulations,

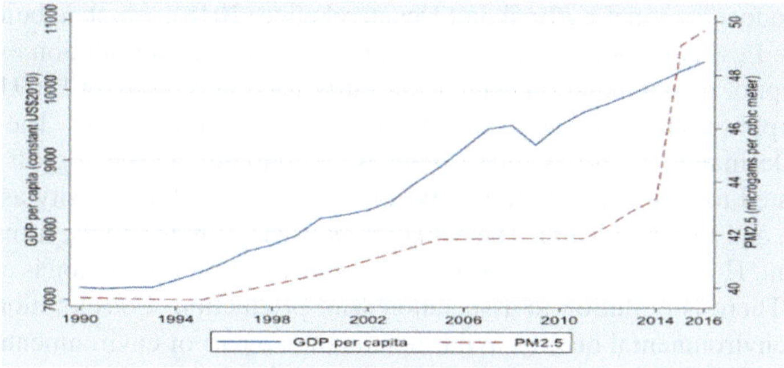

Fig. 3.23 Global trends and GDP and air pollution, 1990–2016

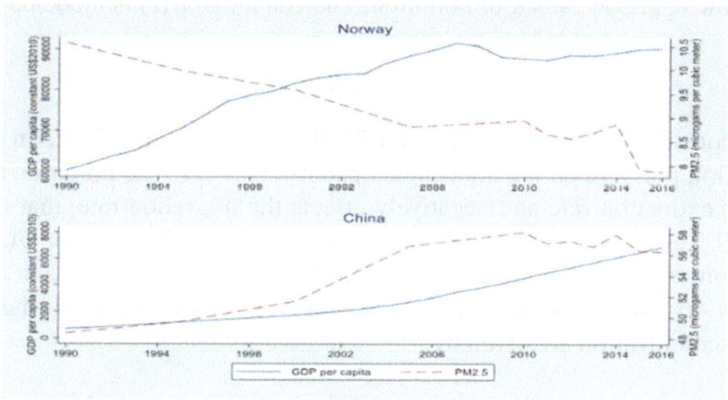

Fig. 3.24 Norway and China show the opposite relations, 1990–2016

controlling coal-burning power plants, and reducing the number of cars, to reduce pollution.

Russell Means points out a source of the concerns: "All European tradition, Marxists included, has conspired to defy the natural order of all things. Mother Earth has been abused, the powers have been abused, and this cannot go on forever. No theory can alter that simple fact. Mother Earth will retaliate, the whole environment will retaliate, and the abusers will be eliminated." The 2021 Nobel Prize in physics was unexpectedly awarded to Syukuro Manabe and Klaus Hasselmann for their works, which laid the foundation of our knowledge of the Earth's climate and how humanity affects it. The Royal Swedish Academy of Sciences chose them for their "groundbreaking contributions to our understanding of complex physical systems" and for building "the physical modelling of Earth's climate, qualifying variability and reliably predicting global warning." The environment is not strongly emphasized in most of the main streams of economics over time. E.O. Wilson says: "If all mankind were to disappear, the world regenerate back to the rich state of equilibrium that existed ten thousand years ago. If insects were to vanish, the environment would collapse into chaos." Man causes collapses of natural order as well as human societies.

I now introduce a model by Bosi and Desmarchelier (2018), which is built by integrating the Levins model and the Solow-Ramsey model with pollution externality. The concept of metapopulation, which was introduced in ecology in 1969 by Levins (1969), implies a spatially fragmented population of the same species. The model by Bosi and Desmarchelier is a synthesis of ideas in economics and ecology. It addresses issues related to coupled evolution between economy and biodiversity as modeled by Lafuite and Loreau (2017). The economy consists of households, firms, and the government. Households work, consume, and enjoy nature. Production is carried out by firms. There is pollution in association with production. Consumption is influenced by environmental quality. Wildlife habitat is a part of environmental quality. A drop in wildlife entails lower consumption. Wildlife is assimilated to a single metapopulation. The government gets income from taxing the firms to improve the environment. Let q stand for the fraction of patches occupied at a given time. Let P represent the aggregate stock of pollution. The change of $q(t)$ is modeled as:

$$\dot{q} = \varphi(P)q(1 - q) - \beta(P)q,$$

where φ represents the migration rate and β is the extinction rate. The term φq means the migration pressure on the unoccupied patches $(1 - q)$. The pollution positively affects the extinction rate and negatively affects the migration rate, that is, $\varphi' < 0$ and $\beta' > 0$. The functions are specified as follows: $\beta = A_\beta P^{\varepsilon_\beta}$ and $\varphi = A_\varphi P^{\varepsilon_\varphi}$. The colonization rate is defined as the difference between the migration and extinction rate: $s \equiv \varphi - \beta$. The pollution elasticities of migration and extinction and the pollution impact on colonization are given by:

$$\varepsilon_\varphi \equiv \frac{P\varphi'}{\varphi}, \varepsilon_\beta \equiv \frac{P\beta'}{\beta}, d \equiv \varepsilon_\varphi - \varepsilon_\beta.$$

The per capita output given by the neoclassical technology is denoted by $f(k) = Ak^\alpha$, where $k = K/L$, K being the total capital stock and L the labor force. Let r and w denote the real interest rate and wage rate, respectively. The marginal conditions for profit maximization are:

$$r = (1 - \tau)f', w = (1 - \tau)(f - kf\prime),$$

where τ is the tax rate.

Let $c(t)$ and $h(t)$ stand for the consumption level and wealth of the representative household. The household maximizes the utility subject to the budget constraint:

$$Max \int_0^\infty \frac{(cq^\eta)^{1-\varepsilon}}{1 - \varepsilon} e^{-\theta t} dt$$

subject to:

$$c + \dot{h} \leq w + (r - \delta)h,$$

in which θ stands for the rate of time preference, δ the depreciation rate of physical capital, η the propensity to wildlife, and $1/\varepsilon$ the elasticity of intertemporal substitution of the composite good (cq^{η}).

The government uses its tax revenue to finance the depollution expenditures M:

$$M = \tau f N.$$

The pollution accumulation follows a linear question:

$$\dot{P} = -aP + bY - \gamma M, a, b, \gamma \geq 0,$$

where a, b, and γ stand for, respectively, the natural rate of pollution absorption, the rate of population accumulation owing to economic production, and the pollution abatement efficiency. Bosi and Desmarchelier choose the following parameter values for numerical simulation:

$$A = 1, A_{\beta} = 1, \varepsilon_{\beta} = 1, A_{\varphi} = 1, \varepsilon_{\varphi} = -1, \alpha = 0.33,$$

$$\theta = 0.01, \delta = 0.025, \tau = 0.002, a = 0.003, b = 0.0015, \gamma = 0.3.$$

There is an unstable limit cycle as in Fig. 3.25 (Bosi and Desmarchelier 2018). Even in the long term, the system will not arrive at an equilibrium point.

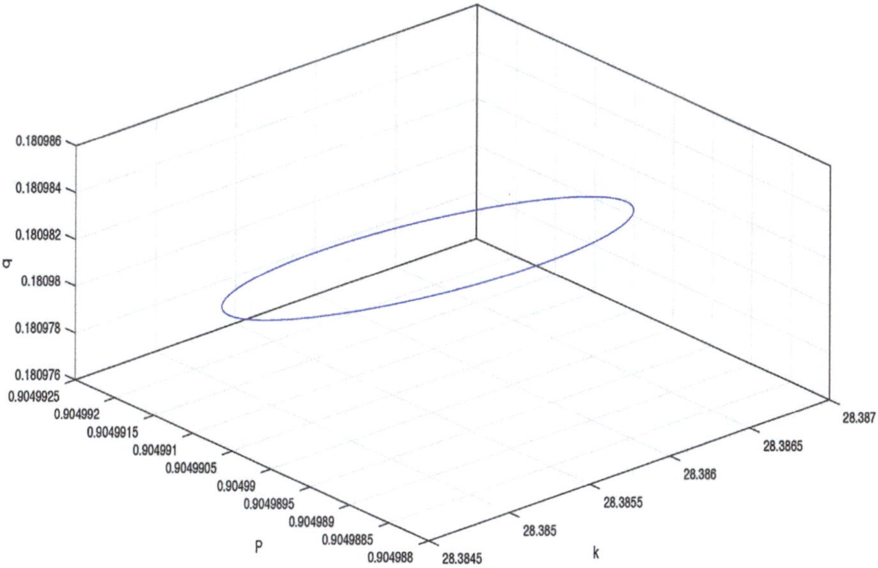

Fig. 3.25 Existence of limit cycles with endogenous environment

References

Bella G., Mattana P, Venturi, B (2017) Shilnikov chaos in the Lucas model of endogenous growth. Journal of Economic Theory 172: 451–77

Benhabib J, Perli R (1994) Uniqueness and Indeterminacy on the Dynamics of Endogenous Growth. Journal of Economic Theory 63:113–42

Bosi S and Desmarchelier D (2018) An economic model of metapopulation dynamics. Ecological Modelling 387: 196–204

Clark CW (1976) Mathematical bioeconomics: The optimal management of renewable resources. Wiley, New York

Clark CW (1985) Bioeconomic modelling and fisheries management. Wiley, New York

Dang HH, Fu HH, Serajudoin U (2020) Does GDP growth necessitate environmental degeneration? Retrieved from: https://blogs.worldbank.org/opendata /does-gdp-growth-necessitate-environmental-degradation

Dikes C, Wang JX (2016) Can a stochastic cusp catastrophe model explain housing market crashes? Journal of Economic Dynamics & Control 69: 68–88

Domar EV (1946) Capital expansion, rate of growth and employment. Econometrica 14: 137–47

Guastello SJ (2001) Nonlinear dynamics in psychology. Discrete Dynamics in Nature and Society 6:11–29

Guevara GC and Escot L (2021) Monetary policy rules: An approach based on the theory of chaos control. Results in Control and Optimization 4: 1–16

Gupta RP, Banerjee M, Chandra P (2014) Period doubling cascades of prey-predator model with nonlinear harvesting and control of over exploitation through taxation. Communications in Nonlinear Science and Numerical Simulation 19: 2382–405

Haavelmo T (1954) A study in the theory of economic evolution. Amsterdam, North-Holland

Harrod RF (1939) An essay in dynamic theory. The Economic Journal 49: 14–33

Harrod RF (1948) Towards a dynamic economics. Macmillan, London

Henderson DW, McKibbin W (1993) A comparison of some basic monetary policy regimes for open economies: Implications of different degrees of instrument adjustment and wage persistence. Carnegie-Rochester Conference Series on Public Policy 39: 221–318

Kaldor N (1940) A model of the trade model. Economic Journal 50: 78–92

Kalecki M (1935) A macrodynamic theory of business cycles. Econometrica 3: 327–44

Krawiec C, Szydlowski M (1999) The Kaldor-Kalecki business cycle model. Annals of Operations Research 89: 89–100

Lafuite AS, Loreau, M (2017) Time-delayed biodiversity feedbacks and the sustainability of social-ecological systems. Ecological Modelling 351: 96–108

Levins R (1969) Some demographic and generic consequences of environmental heterogeneity for biological control. Bulletin of the Entomological Society of America 15: 237–40

Lorenz EN (1963) Deterministic nonperiodic flow. Journal of the Atmospheric Sciences 20: 130–41

Lucas RE (1988) On the mechanics of economic development. Journal of Monetary Economics 22: 3–42

Peng Y, Lu Q, Xiao Y, Wu X (2019) Complex dynamics analysis for a remanufacturing duopoly model with nonlinear costs. Physica A 514: 658–670

Ramsey F (1928) A mathematical theory of saving. Economic Journal 38: 543–59

Solow R (1956) A contribution to the theory of growth. Quarterly Journal of Economics 70: 65–94

Sparrow C (1982) The Lorenz equations, bifurcations, chaos, and strange attractors. Springer, Berlin

Swan TW (1956) Economic growth and capital accumulation. Economic Record XXXII: 334–61

Taylor JB (1993) Discretion versus policy rules in practice. Carnegie-Rochester Conference Series on Public Policy 39: 195–214

Thom R (1972) Structural stability and morphogenesis. Benjamin, New York

Wilson A (2011) Catastrophe theory and bifurcation: Applications to urban and regional systems. Croom Helm, London

Yang XN, Yu P, Yue X, Xue W (2019) Nonlinear dynamics of a duopoly Stackelberg game with marginal costs. Nonlinear Analysis: Real World Applications 39: 185–191
Yu YH, Peng MS (2016) Stability and bifurcation analysis for the Kaldor-Kalecki model with a discrete delay and a distributed delay. Physica A 460: 66–75
Zeeman EC (1977) Catastrophe theory: Selected papers 1972–1977. Addison-Wesley. Reading, MA
Zhang WB (1990) The complexity of nonlinear dynamic economic systems - The Kaldorian model with government policy of bond finance. Journal of Mathematical Sociology 15:15–34
Zhang WB (1991) Synergetic economics. Springer, Heidelberg
Zhang WB (2005) Differential equations, bifurcations, and chaos in economics, in Series on advances in mathematics for applied sciences, vol. 68. World Scientific, Singapore
Zhang WB (2009) Monetary growth theory: Money, interest, prices, capital, knowledge, and economic structure over time and space. Routledge, London
Zhang WB (2020a) The general economic theory: An integrative approach. Springer International Publishing, Switzerland
Zhang WB (2020b) The Taylor rule and business cycles in the Solow-Tobin model. Asian Journal of Economics and Business1: 95–109
Zhang WB (1997) The complexity of economic evolution - change speeds and time scales, in Fang, FK, Sanglier, S (eds.) Complexity and self-organization in social and economic systems. Springer, Berlin
Zhang WB (2003a) Complexity and sustainable development, an article for the theme of *Fundamental economics* in Encyclopedia of Social Sciences and Humanities (http://www.eolss.com), edited by the UNESCO.
Zhao, LW, Du JG, Wang QW (2019) Nonlinear analysis and chaos control of the complex dynamics of multi-market Cournot game with bounded rationality. Mathematics and Computer in Simulation 162: 45–57

Chapter 4
Self-Organization, Catastrophes, and Structural Changes

In his letter to Oldenberg in 1675, Isaac Newton writes: "The power of life and will by which animals move their bodies with great and lasting force … demonstrate that there have to be other (undiscovered) laws of motion." In his *Matter and Motion* in 1877, James. C. Maxwell describes:

> Physical science is that department of knowledge which relates to the order of nature… Nature is applied to those branches of science in which the phenomena considered are of the simplest and abstract kind, excluding the consideration of the more complex kind, such as those observed in living things.

Traditional scientists think of the complexity of living things but do not provide some sophisticated (mathematical) theories to reveal the secrets of life. When it comes to man and society, "The term self-organization refers to the role of the self-conscious, creative, reflective and knowledgeable human beings in the reproduction of social systems." (Fuchs 2003, p. 16). As a complex system cannot be understood merely by breaking it down into multiple parts, the traditional scientific approach cannot explain properly the behavior of complex systems. Interdependence between parts creates phenomena that cannot be observed by any part in isolation. Structural changes are not exceptional in the literature about humanity, love, and the history of social evolution. History is always unpredictable, even though it is often constructed as it appears certain. Harvey (2001) argues:

> … social structures, unlike natural structures, are irreversibly evolving constellations…. [S]ocial structures and institutions … have evolutionary life histories of their own: they are born, are subject to temporal cycles of growth and decay, and eventually cast into the dustbin of history. In both instances, short-term and long-term, social institutions are space-time dependent in their ability to influence human conduct at any given time and place (p. 171).

Von der Malsburg (1997) explains the emergence of consciousness:

> Science has always had its greatest triumph when and where it succeeded in subdividing complex phenomena into very simple paradigms. … None of the isolated components of the brain can be expected to hold the essence of consciousness. That resides in the modes of interaction of all parts of the brain, and maybe even in the way the brain is integrated into the social world and the world at large.

W. Zhang, *Complexity Theory and Uncertainties*, Understanding Complex Systems,
https://doi.org/10.1007/978-3-031-42394-9_4

Complexity, defined by Gell-Mann (1994), is the length of the schema required to describe and predict the properties of an incoming data stream by identifying its regularities. A human being is an organic complexity. A person constructs life by self-organization and interacting with the environment. There are strong interconnections among various aspects and functions within the human body (e.g., Haken et al. 1985; Kelso 1995). The connectivity and complexity should be the essence of the study of man and human nature.

4.1 Self-organizing Ant Societies and Collective Survival

Nikola Tesla tells his opinion of life: "Life is and will ever remain an equation incapable of solution, but it contains certain known factors." Life is not an unsolvable equation; life is a process of creating equations. Even if it contains a few known factors, life would remain unpredictable. How human races and species survive, not to mention how they survive well, is one of the most challenging questions. Sustainability or capacity of survival is different from being strong and being successful in daily use senses. Survival of the fittest describes the mechanism of natural selection. In biology, it refers to reproductive success, measured in the population size over time. In Darwin's time, many scientific terminologies were not yet available. Darwin used the term copies itself to describe fitness. One well-mentioned phenomenon of evolution, originally claimed by scientists Edward O. Wilson and Bert Hoelldobler, is that the weight of all the ants in the world has been estimated to be as much as all human beings. To what degree this estimation is reliable is still in debate. Nevertheless, from self-organization perspectives, ant societies, like many other worlds of some other animals, like dogs, cats, monkeys, tigers, and wolves, provide insights into the complexity of survival fitness.

Ants are one of the most abundant and successful species on Earth. They live in complex and cooperative societies within well-constructed elaborate nests. Ant societies are far more sophisticated and orderly than many human societies, even in the contemporary world. They live diligently, prosperously, and orderly. They are affected and make impacts on the ecosystems they live in. They disperse seeds and consume plants. Society is characterized by a sterile caste (workers) and a dedicated reproductive caste (the queen or queen). Ants predate other insect species. They perform effective division of labor some million years earlier than Adam Smith has a systematical explanation of the division of labor and efficiency of production.

Ant evolves into different kinds according to the natural environment and self-organization mechanisms. The combination of inner nature and environment leads to a complexity of the art world which man has begun to know more about. Each ant's behavior, like each human, is seemingly chaotic and unpredictable. But ant society displays a highly ordered structure and the ant, different from wolf and tiger, has survived great numbers of all-natural disasters and environmental catastrophes caused by human progress.

Ant is exemplified by the collective behavior with diligence, “dutifulness,” and mindless drudge, without resting and reflecting. Extensive studies of ant societies over many years by Gordon and other researchers have revealed the complexity of ants with great similarity to the human race from many perspectives. Ants conduct production, reproduction, food supply, division of labor, division of sex, distribution of consumption, task allocation, and maintain social order and stability of the society. Ants, like some bees, wasps, and termites, are eusocial in the sense that reproduction is dominated by one or a few queens and most of the other members never reproduce but carry out jobs such as foraging, constructing, baby-caring, cleaning, and defending. Evolutionary biologists initially were puzzled by the phenomenon that workers do not continue their genes. W. D. Hamilton develops a theory of inclusive fitness and kin selection which argues that non-reproductive workers help to ensure the survival of their siblings. The concept of inclusive fitness measures the total reproductive success of an individual, including direct fitness (one’s offspring) and indirective fitness (gained by helping relatives to pass genes). Siblings share a large percentage of their genetic information. Ant complete jobs and behave collectively orderly by accident, by experimenting with and constantly testing environments in searching for food and task allocation. Ants do not have a managerial system with some leaders in each section of the organization, as inhuman societies. Like many complex systems, a combination of simple individual behaviors generates complicated global phenomena. Ants can collectively become strong destructive forces for human buildings. A Chinese proverb says, ants, small as they are, can damage a great dam.

In some kinds of ants, many thousands of virgin queens and males leave the nest in a very short time, searching for a mate from another colony. Few can survive before they establish their colonies. Predators such as defensive workers of other ant colonies, birds, beetles, spiders, and frogs, will end their lives. Ants pass complete metamorphosis, starting from egg to larva, to pupa, to adult. Ant colony starts with the queen. They live in colonies reaching up to some million members. They work together to gather food and care for the young in highly methodical and coordinated manners. Ant behaves depending on the ant’s caste. The colony is a self-organizing system with a well-defined division of labor and consumption between the castes (e.g., Gordon 1999, 2010). The ants who work outside the nest carry out foraging, patrolling, nest maintenance, midden work, and so on. The ants who work inside the nest are the reserve, seed processor, brood caregiver, and queen. There is no direct authority over the behavior of the other roles. Queens are females who are fed when they are larvae. They lay all the eggs in the colony. Queens have wings and fly to get a mate. They tear their wings off when creating a colony. After mating, the queen has a colony (nest) and raises her first worker offspring. The queen may copulate with a few males in a brief mating period. The queen starts to search for a site for her new nest. She digs a nest, lays eggs, and single-handedly brings up her first babies who are solely female workers. After the period, she never mates again. She stores sperm in a pouch near the tip of her abdomen. The sperm remains immobile until she opens a valve to allow the sperms to its reproductive track for fertilizing eggs. After the first offspring, she never works except for laying eggs. The queen controls the sex of her offspring. Fertilized eggs produce females, either wingless non-reproductive workers or reproductive virgin queens. Virgin queens are produced only when the colony has sufficient workers to allow for its expansion. Unfertilized eggs develop into winged males who do not work and cannot feed themselves. Workers are females and don't reproduce. They do all the jobs in the colony. Males leave the nest to reproduce. They mate with queens. They die shortly after mating because they don't know how to feed themselves.

The initial phase of colony development is the founding stage. Then, the young daughters will take care of the queen. They enlarge the nest, construct sophisticated tunnel systems, and transport new eggs into hatching chambers. The workers feed, clean, and protect larvae till the larvae become workers. Ants don't have good vision. Their first interactions are locally conducted by touching antennae, using their sense of smell to assess ants that come across their way. The touch informs whether the ant is a threat from a rival colony or what the ant is currently doing. Ants communicate using chemical scents called pheromones. Ants like to go along the trails that have the strongest pheromone scent. Ants traveling on a shorter trail between the nest and food make more trips than those along longer trails, leaving more pheromones.

Task allocation is the process that organizes the number of workers in each task. Task allocation is adaptable to the environment. Ant society's task allocation is operated neither by a central authority nor by a hierarchical system. The number of workers in each task varies according to the environment. Weather, food availability, threats, and other environmental conditions all affect worker distribution. Ants can choose tasks according to the need of jobs and ants are "aware" of how many ants are already engaged in the job. Ants can do this not because they have "global" views or information but because they make local communications. An ant's decision on a task is influenced by what he encounters in the environment and the rate of interaction with ants performing different tasks. An ant can be informed about the task another ant has been doing by chemical residues. Ants behave like "statisticians." The size of the colony affects decision-makings of statisticians in task allocation.

When ants are foraging, they move randomly in many different directions. When an ant encounters a food resource, it returns to the nest, leaving a pheromone (a chemical that attracts other ants) trail. Ants who encounter the trail will follow it to find the food source. The pheromone dissipates if there is no enforcement. Ants will enforce the trail if they continue to find food. When there are multiple trails for the same source, ants tend to collectively choose the shortest trail. When there are multiple food sources, at any given time the trails and their strengths encode the colony's collective information about environmental conditions. The information varies as the environment changes.

The forager ant drops the seed at the entrance tunnel to the nest. The food is transported by other ants into the nest. The forger ant continues to search for food. Some ants conserve their resources. Ant society is self-organized via cooperation. Ants are so cooperative that some ants will not do some tasks to serve the colony. Ants are collectively adaptive in the sense that, for instance, their reaching efforts are dependent availability of food. Some colonies keep some ants "unemployed."

Some ant societies keep reserve ants who are idle for some time. "The unemployment" is collectively not inefficient. These ants are not lazy and unskilled. They have a passion to work. When there are some emergent needs, they soon join the work army. The reserve allows the colony to have a faster response time to sudden environmental changes. Ants learn that it would benefit the colony to have some members unoccupied. The colony experiences a stage of development. The initial stage is characterized by work and expansion. It takes some time before the colony becomes large enough to start the reproductive stage when virgin queens and males are produced and leave the nest to begin the cycle anew. The death of the queen implies the end of the colony, except for some species of ants whose nest may have multiple queens.

The complexity of ant societies, like human societies, is examined with conceptions of structures, nonlinear interactions, and networking. Green (2014) compares humans with ants. Just as ants create an anthill without being aware of it, human activities include many unintended side effects, which cause unforeseen negative and crises. Green traces how these trends emerge. Social and economic systems are suffering from many unpredictable side effects, such as environmental disasters and climate changes, of modernization. Beeren (2012) studies how ant colonies are parasitized simultaneously by several arthropod species. In a study of the complexity of ant societies, the relationship between colony size and ant division of labor is another important topic. Increases in colony size tend to increase the complexity of the division of labor. Ferguson-Gow et al. (2014) studied the Attine ants, a large neotropical group including the leaf-cutter ants. When eusociality initially evolved, colonies might be small, and the worker caste might not entirely refrain from reproduction. In small colonies, a worker, if she is not sterile, has greater chances (than in larger colonies) to take over the colony. As the colony size is enlarged because of evolution, the specification becomes more complicated. In the biology of evolution, the size-complexity hypothesis refers to that as colony size increases, workers and queens maximize their inclusive fitness by the specification. An individual worker has little chance to become the queen. A worker might have a higher probability to maximize indirect fitness in large colonies in her role. Increases in the division of labor tend to increase collective efficiency, which would make the colony more successful and further enlarge the colony. There are positive feedback loops between size, division of labor, specification, and colony efficiency. There are enormous colonies of millions of workers in some ant species. Nooten et al. (2019) study how habitual complexity affects functional traits and diversity of ant assemblages. Species within habitat complexity conferred by vegetation characteristics persist only if their functional traits are matched to the environment. They measure key functional traits of ants, such as body size, femur length, antenna scape length, and head length/head. They find that in wooded habitats, ants are larger and have broader heads. While in habitats with dense herb/grass layers ants with longer antenna scape prevail. Their study shows that vegetation structural complexity can drive ant assemblages in terms of both species numbers and functional traits.

Ant societies have tragedies of the commons. For instance, when they get some wrong signals, millions of ants can be addicted to moving toward a single central point. They collectively suicide. Like human overfishing, an ant colony may destroy its environment to survive through exploitation. The ants are obsessed with their collective self-destructive behavior for millions of years without changing behavioral patterns.

4.2 The Chinese Ancient Classics *Yi Jing* and Chaos Theory

Complexity theory provides a new and broad vision of change. It also provides a new framework for studying world views of different cultures. I now apply complexity theory to the *Yi Jing* (the *Book of Change*, or the *I Ching*). It is an ancient oracle, a divinatory book that basically "determined" the Chinese vision of nature, man, and society. Comprehension of the *Yi Jing* is the key to approaching cultures, not only of China, but also overseas Chinese societies, Japan, Korea, and Vietnam. The book was officially used as a classic for the elite to digest and interpret before the Western way of thought and thinking way penetrated and dominated these Confucian regions. The essential idea is that everything is changeable, and change is owing to the interdependence between two basic elements, called *yin* and *yang*. There is the sun (light) and the moon (darkness)—a pair from the same universe. There are chickens and eggs because there is interdependence between the two. The relative position of the sun and the moon, plus other *yin–yang* combinations—such as winds, clouds, emotions, mountains, rivers, flowers, and birds—are the sources of unaccountable poems in the *yin–yang* region. Affairs move between the two poles. Motion is unpredictable but there are limits or boundaries to change. Perfection is dead and chaos is the manner toward perfection. Order and disorder coexist. Capitalism is unstable but effective, while socialism is stabilized but dying. Chinese philosophy does believe in the origin of life, but the origin is chaotic or unknown to humans. There is Dao. Dao is referred to how *yin* and *yang* interact or the law of nature in contemporary terminology.

Affairs are neither random nor rationally deterministic, according to *Yi Jing*. It is timing that combines randomness and deterministic forces. The organic system thus depends on timing. Without timing, affairs cannot be appreciated, and action cannot be proper. Meaningfulness occurs in time and space. A faith diffused over entire humanity is never perceived in the Chinese way of thinking. Christianity was known to China more than a thousand years ago. It was known to Korea and Japan a long time ago. Religion was not spread culturally, spiritually, or intellectually because of, except conflicts of interests and powers, the *Yi Jing* vision of the world. It was only after the West had dominated the world that these Confucian regions began to "accept" Western religions. In his comments on the *Yi Jing*, Jung (Sabadini 2019; Jung 1950) writes:

> The Chinese mind, as I see it at work in the I Ching, seems to be exclusively preoccupied with the chance aspect of events....
>
> The matter of interest seems to be the configuration formed by chance events in the moment of observation, and not at all the hypothetical reasons that seemingly account for the coincidence. While the Western mind carefully sifts, weighs, selects, classifies, isolates,

the Chinese picture of the moment encompasses everything down to the minutest nonsensical detail, because all the ingredients make up the observed moment….

This assumption involves a certain curious principle that I have termed synchronicity, a concept that formulates a point of view diametrically opposed to that of causality... Synchronicity takes the coincidence of events in space and time as meaning something more than mere chance, namely, a peculiar interdependence of objective events among themselves as well as with the subjective (psychic) state of the observer or observers.

The dominant way of Chinese thinking was described by Jullien (1995):

The fact that there is an overlapping of different domains in Chinese thought reveals a common model that runs through the entire culture, namely, of a configuration or disposition of things operating through opposition and correlation, and which constitutes a working system (p. 17).

What Jullien portrayed is the *Yi Ying*-framed way of Chinese thinking. Jung (1997) observed:

The type of thinking based on the synchronistic principle, which reached its climax in the *Yi Jing*, is the purest expression of Chinese thinking in general. In the West it has been absent from the history of philosophy since the time of Heraclitus and reappears only as a faint echo in Leibniz. However, in the interim, it was not altogether extinguished but lingered on in the twilight of astrological speculation, and it remains on that level today (p. 85).

The texts of the *Yi Jing* consist of sixty-four hexagrams and judgments on them. The hexagrams were supposed to symbolize all situations. Each hexagram is composed of linear signs, the judgment of the whole hexagram, and the text about the individual lines. Each hexagram is supposed to represent one or more phenomena of nature or society or individual. The discovery of the eight trigrams (three-line symbols) is attributed to the first of the legendary five emperors, Fu Xi (ruled 2852–2737 B.C.). The original eight trigrams did not have philosophical significance until the Ten Appendices were added to them, and subsequently, the eight trigrams were extended to the sixty-four hexagrams. Fu Xi designated an unbroken (——) and a broken (– –) line as elementary symbols for the universe. The two lines and their symbolic significance were devised by the emperor. The unbroken line stands for *yang*, which means man, positiveness, progress, clarity, strength, and light. The broken line represents *yin*, which means woman, negativeness, declination, obscurity, weakness, and darkness. Thus, the *yin* and *yang* are used to express the two extremes or poles of universal phenomena. The universe came to be a result of the interactions between the two opposing universal forces. Using the two lines, Fu Xi invented the eight trigrams. Each trigram may represent a state of phenomena under consideration. The eight trigrams show only eight patterns of phenomena. In Fig. 1.1, we illustrated the difference between Newtonian vision (also reflected in the theories on man and societies by Adam Smith 1759, 1776) and the *yin–yang* world vision with the eight trigrams. It is only in the last few decades that the spread of complexity theory in natural and social sciences equipped with computers has made Chinese philosophies easier to understand for scientifically trained minds. It is easy to see why Newtonian-trained scholars tend to think of Chinese culture as a mystery

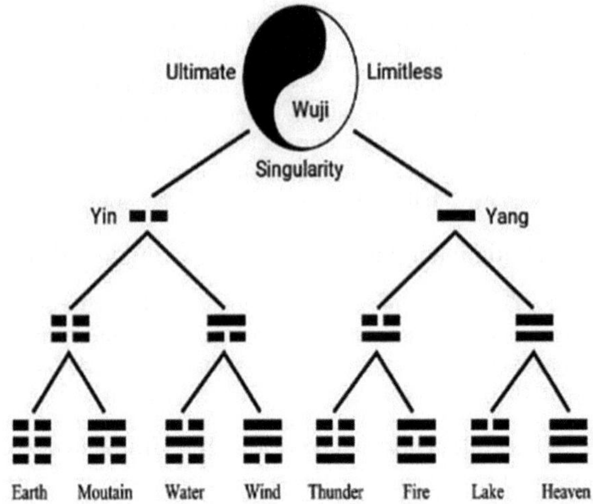

Fig. 4.1 Infinite variety born from the unidentifiable one. *Source* https://www.chinasage.info/yi-jing/i-ching.htm

and irrational as the Newtonian frame visualizes the world differently from the *Yi Jing*-based one.

A further extension by combining the eight trigrams in pairs results in the sixty-four hexagrams. There is no general agreement about when the eight trigrams were combined to form the sixty-four hexagrams. It was alleged that it was King Wen (r. 1171–1122 B.C.) who translated the symbols of the eight trigrams into words. During the seven years when he was held under house arrest, he named the sixty-four hexagrams and arranged them in sequence. It has been maintained that the two texts, the *Judgment,* and the text of the individual lines of the sixty-four hexagrams were given by King Wen or Duke Zhou (d. 1094 B.C.). Figure 4.1 illustrates various patterns of the *yin–yang* system.

I can illustrate the vision of the *yin–yang* pair dynamics with chaos theory. A one-dimensional map can be used to generally illustrate the Chinese mentality and Chinese world vision. If I locate the *yin* at the origin and the *yang* at the unity and assume that the time-dependent state variable moves between the *yin* and *yang* by the law described by the logistic map in which $x(t)$ in period t stands for the distance from the *yin* and $1-x(t)$ in t for the distance from the *yang*, I get some hints about the Chinese vision about the world structure from the understanding the simple logistic model. The logistic map, proposed by Robert May in 1976, shows the state of the *yin* in period $t+1$ is given as follows: $x(t+1) = rx(t)[1-x(t)]$. Tomorrow's *yin* power is dependent on today's *yin* power and today's *yang* power under a given environment measured by r. How the logistical map and its various extensions can be applied to explore hidden implications of the *Yi Jing* have been studied by different authors in various fields. The simple *yin–yang*-pair interaction can lead to unlimited complicated phenomena as portrayed in Fig. 4.2. For instance, interactions between male and female powers under various physical, cultural, social, and economic conditions lead to unlimited love versus hate as well as trust versus deception stories across

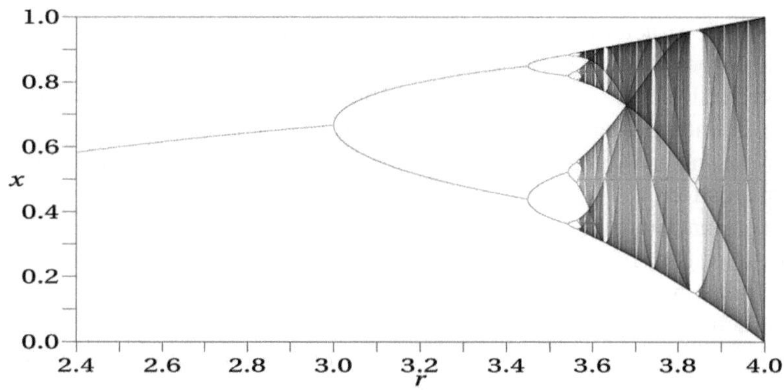

Fig. 4.2 Complexity of a *yin–yang* interdependence in different environments. *Source* https://en.wikipedia.org/wiki/Logistic_map

races. The dynamic force is simple, even though its hidden behavior could have only recently been revealed owing to the computer which was available not long time ago. The logistic map exhibits many types of long-term behavior that can be displayed by a dynamic system: stable and unstable fixed points, periodic orbits, aperiodic trajectories, and chaos. Here, random shocks are not required to produce chaos. One of the fundamental aspects of chaos is that many different possible motions are simultaneously present in the system. There is typically an infinite number of unstable periodic orbits that coexist with the chaotic motion. Because such "chaotic" trajectories are unstable, errors of estimation in parameters or initial conditions, however tiny, will accumulate rapidly into substantial errors in the forecast. The characteristic of extreme sensibility to tiny perturbations is known as the butterfly effect. The future behavior of such a chaotic model solution cannot be anticipated from its patterns in the past.

4.3 Cultural and Institutional Bifurcations Before Industrialization in Europe and China

Human relations and cultural patterns are through conscious construction under natural conditions. Geography plays an essential role in cultural evolution. Social organizations evolve for people to collectively survive. A traditional tribe and a modern state have different levels of organizational complexity. Globalization implies also global adaptions and innovations of institutional and organizational structures. It might take a few generations for mankind to achieve a relatively stable global cultural, political, and economic equilibrium. Man is tool based. Technology fundamentally frames men's relations with nature and among men. Technological change is an integrated part of human evolution. Technology brings about social and economic structural changes. Human institutions, moral codes, and formal or informal enforcement of rules are closely related to technology. Industrialization is characterized by technological changes and human institutional and moral adaptions. Modern technology, modern mathematics, and natural sciences are mainly created in the West and are accepted as universal truths. Both socialism and capitalism are Western creations.

China is an interesting example of how a culture integrates elements of other cultures into its society. Greif and Tabellini (2010) study the bifurcation of social organizations in premodern China and Europe. Their study focuses on the distinct social structures and cities that sustain cooperation in their societies. Chinese social structure is kinship-based. The family organization is hierarchical and is sustained by moral codes and reputation among clan members. In Medieval Europe, the city was a key cooperative organization and kinship lines, and external enforcement of rules played a big role. The clan is a kinship-based community with which its members identify. Cooperation within the clan is maintained mainly by habitual and moral obligations and respected persons. Conflicts resulting from, for instance, cheating and free riding, are often solved not through formal institutions. The city is composed of many clans and cooperation is mainly sustained through the formal enforcement of rules. Morality does not have a major role. The clan is not costly in terms of legal operations and enforcements, whereas the city makes it easier for heterogeneous agents to effectively cooperate. Clan culture is likely to be formed in a society dominated by clan loyalty, whereas in a society with city culture, moral obligations have a wider scope. Members of a clan-dominated society are cooperative with clan-limited members, whereas members of a city-based society are cooperative with generalized morality and formal institutions. There are interactions between moral formation, organizational structuring, and social and economic efficiencies.

By 1000 CE, large kinship organizations were common in China but not in Europe owing to geographical, religious, and other factors. I could take the period around 1000 CE as the initial conditions for the two cultures and examine development onward. Fei and Liu (1982) explain the survival of kinship structures in China: "the clan as a Chinese institution in the pre-modern period … prevailed some 800 years, beginning with the Sung dynasty." In Europe, tribal relationships were gradually weakened by the Church, and generalized morality was strengthened. The Church discouraged some kinship relationships, which might strengthen kinships, such as polygamy, marriages without the women's content, concubinages, and marriage among distant kins. By the ninth century, the nuclear family was dominant. Legal codes no longer connect rights and kinship. Large kinship groups were marginal. With regards to China, by 1000, organizations based on kinship were rare and generalized morality was emphasized. As described by Buckley and Watson (1986), China had been "the predominant form of kinship organizations in late imperial China." Education, religious services, relief from poverty, local public goods, and many other human relations were carried out within kinship organizations. The state, dominated by the strongest clan, the emperor family, controlled clans by rules and promoting beliefs (such as neo-Confucianism). Intraclan enforcement left little place for formal enforcement institutions. A legal system based on generalized morality would undermine the clans. The Chinese state needed intraclan dispute resolution because it enabled the emperor's family to continue controlling the huge empire.

A high civilization is characterized by a high variety of thoughts, ideas, spirits, expressions of emotions, and dimensions of freedom. Different from Europe, China's social order was mainly under the emperor (the family and associated obedient bureaucrats) and clan-based collections of villages. Clan-based China did not effectively develop urbanization and self-government. The whole country was maintained for the survival of the emperor family domination with multiple levels of clans. China's urbanization rate was between 3 and 4% during the period of the eleventh to the nineteenth century, whereas Europe's urbanization was initially also low but rose about 10%. China's small cities were venues for cooperation among members of local clans rather than the melting of heterogenous households. European cities were characterized by self-governance. Friedmann (2007) described the situation by the seventeenth century as follows:

> the majority of a city's population consisted of so-called sojourners, people who had come from elsewhere and were considered (and thought of themselves as) only temporary residents … suspicions were always rife that sojourners could not be trusted (p. 274).

The explanation according to kinship is not deniable because it occurred within China. Nevertheless, a society ruled by a dictator can easily be manipulated and controlled by divide and rule. Today, the Chinese government is still applying the same strategy to build a controllable (in the official slogan, harmonious) society by not permitting farmers to become urban citizens. Some years ago, it was far more difficult for mainland-born and educated Chinese to become American or some advanced nations' citizens than to get Shanghai or Beijing urban citizenship (*hukou*). Guild-like organizations (huiguan) were also composed of members from the same place of origin, whereas in Europe individuals created cities with the support of the Church and secular rulers. In Western Europe, immigrants could relatively easily integrate with the city population, and by 1350, cities were mainly maintained by self-governance. Legal infrastructures and legal professionals were developed in association with high crime rates and a high number of policemen per capita in Europe.

Since industrialization took place till the beginning of WWII, Europe and China had experienced structural changes in institutions, cultures, economies, and technologies in association with the rapid growth of rational and scientific knowledge. The rise of the West challenged traditional Chinese society from all aspects. Nevertheless, there is something almost invariant or slowly changed in cultural adaption. Chinese societies are still emphasizing kinship groups more than Western societies. Redding (1993: 66) describes China: "you trust your family absolutely, your friends and acquaintances to the degree that mutual dependence has been established…. With everybody else you make no assumptions about their good will."

4.4 Global Political, Social, and Economic Catastrophes in the 17th Century

Historians call the mid-seventeenth century "the general crisis" because the period experienced more state breakdowns on the Earth than any previous and subsequent ages. Parker (2008) describes:

> In the 1640s. Ming China, the most populous state in the world, collapsed; the Polish-Lithuanian Commonwealth, the largest state in Europe, disintegrated; much of the Spanish monarchy, the first global empire in history, seceded; and the empire Stuart monarchy rebelled - Scotland, Ireland, England, and its American colonies.

Even in the year 1648, for instance, a tide of urban rebellions started in Russia, which was the largest state in the world and the Fronde Revolt almost ruined France, which was the most populous state in Europe. At the same time, there were social revolts and conflicts. Parker (2008) shows:

> in Istanbul (Europe's largest city) irate subjects strangled Sultan Ibrahim, and in London, King Charles I went on trial for war crimes... In the 1650s, Sweden and Demark came close to revolution; Scotland and Ireland disappeared as autonomous states; the Dutch Republic radically changed its form of government; and the Mughal Empire, then the richest state in the world, experienced two years of civil war.

In the mid-seventeenth century, the world experienced many wars. Historians gave varied reasons for these catastrophes and crises. War from personal levels to states is a common way to solve conflicts. The identity and faith of a tribe or state are often a human way of justifying wars against others. Nevertheless, one might ask why the war was so frequently and commonly conducted to solve domestic and international problems in the seventeenth century. Parker (2008) argues that there might be two main causes for the general crisis. One is that population densities in the conflict regions were very high. This implies that people did not have sufficient food for survival. This is one reason for Malthusian poverty traps. The other main cause is extreme weather. Climate change was considered the main factor for simultaneous catastrophes in countries that were not connected. Fewer sunspots and more volcanic activity reduced solar energies received on Earth. Global temperature and climate were changed. Climate change led to living conditions and food supply. Sixteen-forty saw many prolonged droughts in many areas. For instance, in China, famines occurred across populous regions. People died in millions owing to a shortage of food and social instabilities. Destructive peasant rebellions weakened the foundation of the already declining dynasty. The dying Ming dynasty was characterized by widespread corruption, an incapable state from the top to the bottom, and enlarged inequality between the powerful and the poor. A neighboring race, Manchus, which was less than one percent of China's population, took over China in 1664 and governed China till 1911. The Qing dynasty was prosperous with its first three emperors and then started to experience the "standard" declination of a great empire. From the 1800s, the Chinese ruling group, different from the Japanese, resisted the introduction of Western thought for the race's survival with easily manipulated obedient Han Chinese. China began to suffer from one after another foreign humiliation. It was in 1949 that another peasant rebellion leader, Mao, built a new China. The Chinese peasant who neither received education in modern thought nor experienced modern life had nothing to rely on but the traditional Manchu way to rule obedient Chinese with effective modern weapons. After Mao died, Deng who had lived in France for many years, started to modernize China's economy with the traditional Chinese bureaucratic (elite-dominated) structure sustained by the national best-talented people selected through examination systems.

Although there is not sufficient systematical and accurate evidence to prove how strongly global climate change in the mid-seventeenth century caused global catastrophes in political economies, the environmental change does seem to contribute to simultaneous calamities in mutually isolated regions. Modern technologies have enabled humans to affect the global environment simultaneously and strongly. Even if the sun and the Earth move regularly, human action can cause global environmental catastrophes. The sum of many smart people may turn out to be stupid humanity, as human history repeatedly proves. An obvious contemporary example is that a free and highly educated race might collectively perish simply because population reproduction is not fast enough for the race continuation. Race's intellectual contribution will enable some contemporary stupid and lazy ones to enjoy themselves in the future.

4.5 Poverty Trap of Mainland China and Fast Growth of Taiwan From 1949 to 1978

Each society is a complexity made of many levels of complexities. A society is composed of individuals. The sum of individuals' motivations may not mirror the collective behavior of thought of society. The whole is seldom equal to the sum of parts, it is greater or smaller, more virtuous or more evils than the sum of individuals. Even if one is familiar with the thought and values of everyone in a society, one may still be incapable of comprehending collective behavior. A typical example is why Germany behaved collectively so cruelly in WWII, even though Germany had so many enlightened thinkers, excellent philosophers, great scientists, artists, musicians, good citizens, and Nobel laureates in different fields (more than the sum of British and American ones) just before WWII. Many articles and books have been published on the phenomenon, but collective behavior remains a mystery. Economic theory for each level of complexity shows the possible consequences of various mechanisms under various environments. As a new structure, which is the basis for building relations among individuals, emerges in the process of evolution, the theory, which was once valid for the society or validified in other societies is no longer applicable to new situations. A society can hardly succeed by imitating or copying another society. There are a lot of poor countries in the world. Poverty is a consequence of collective inefficiency. A typical case is China. A collective application of a trivial idea in modern economics—market mechanism—has enabled many millions of people to escape from poverty.

China has been experiencing economic miracles since 1978. Nevertheless, from 1949 to the economic reform in 1978, China had been in a poverty trap characterized by low-income per capita. In contrast, Taiwan, which was economically and politically separated from mainland China in 1949, experienced fast economic growth when the mainland was suffering from a poverty trap. By 1949, Taiwan differed from mainland China mainly in that it had taken a lot of human capital (elite) and wealth from the mainland (Zhang 2003b, 2021a). We now introduce an economic model to provide some insights into the bifurcation of the two economies.

I proposed the model when thinking of China's economic growth many years ago. The main factors of economic growth, except institutions and the size of the population, are human capital and physical capital. One of the first seminal attempts to render technical progress endogenous in growth models was initiated by Arrow (1962). He emphasized one aspect of knowledge accumulation—learning by doing. Uzawa (1965) introduced a sector specifying creating knowledge into growth theory. The education sector utilizes labor and the existing stock of knowledge to produce new knowledge, which enhances the productivity of the production sector. I (Zhang 2006, Sect. 3.2) synthesize Solow's physical capital accumulation, Arrow's learning-by-doing, and Uzawa's learning-by-education growth models within a comprehensive framework. It provides another economic mechanism of path-dependent economic development. We now consider an economy that has one production sector, denoted by i, one education sector, denoted by e, and a homogenous and fixed national labor force, N. The labor force is engaged in economic activities, teaching, and studying. The commodity is selected to serve as a numeraire.

Let $F(t)$, $K(t)$, and $H(t)$ stand for, respectively, the output level of the production sector, the national capital stocks, and the level of the human capital of the population. Market-determined wage rate and rate of interest are respectively $w(t)$ and $r(t)$. I express labor force and capital stocks employed by sector j, respectively, N_j and K_j. Production is to combine qualified labor force, $H^m N_i$, and physical capital, K_i. The production process is described by

$$F = AK_i^{\alpha}\left(H^m N_i\right)^{\beta},\ A, \alpha, \beta > 0, \alpha + \beta = 1.$$

The conditions for profit maximization imply:

$$r = (1-\tau)\alpha A k_i^{-\beta},\ w = (1-\tau)\beta A H^m k_i^{\alpha},$$

where $k_i \equiv K_i / H^m N_i$ and τ is the tax rate on the product level. I denote per capita wealth by k, where $k \equiv K/N$. Per capita current income is given by $y = rk + w$. The per capita *disposable income* is given by $\widehat{y}(t) = y(t) + k(t)$. The representative household Maximizes a utility function:

$$U = c^{\xi} s^{\lambda},\ \xi, \lambda > 0, \xi + \lambda = 1$$

subject to:

$$c + s = \hat{y}.$$

The optimal solution: $c = \xi \hat{y}$ and $s = \lambda \hat{y}$. The change in the household's wealth is equal to the savings minus the wealth sold at the time t, i.e.,

$$\dot{k} = s(\hat{y}) - k = \lambda \hat{y} - k.$$

Human capital accumulates through Arrow's learning-by-doing and Uzawa's learning-by-education:

$$\dot{H} = \frac{\upsilon_e K_e^{\alpha_e} (H^m N_\upsilon)^{\beta_\upsilon} (H^m N_e)^{\beta_e}}{N} + \frac{\upsilon_i F}{N H^\pi} - \delta_h H,$$

where N_e is the number of students, $\delta_h (> 0)$ is the depreciation rate of human capital, υ_e, υ_i, α_e, β_υ, and β_e are non-negative parameters. It is assumed that the students and teachers are paid by the government's tax income at the same wage rate as the wage rate of workers. It is assumed that the economy has a fixed ratio of the population who are getting an education in the university: $N_e = n_e N$. The total tax income is used for paying the students, teachers, and capital stocks employed by the university. The government spends $w N_e$ amount of money on students. The time distribution is given by $T_i + T_e = T$, where T is the fixed available time, T_i is the work time and T_e is the time as a student. The student gets free education but does not receive any wage. The budget for paying teachers and the capital stocks of the university is satisfied: $w N_\upsilon + r K_e = \tau F - w N_e$.

The university distributes its total resource $\tau F - w N_e$ to the teachers N_υ and the capital stocks K_e in such a way that the output of the university will be maximized:

$$\text{Max} \upsilon_e K_e^{\alpha_e} \left(H^m N_\upsilon\right)^{\beta_\upsilon} \left(H^m N_e\right)^{\beta_e}$$

subject to the budget.

Labor force and capital stocks are fully employed.

I now examine multiple equilibrium points by simulation. The parameters are specified as follows:

$$\alpha = 0.35,\ N = 1,\ A = 2,\ N_e = 0.06,\ \tau = 0.08,\ \lambda = 0.7,\ \alpha_e = 0.7,$$
$$\beta_e = 0.7,\ \beta_\upsilon = 0.7,\ \nu_e = 1.8,\ \nu_i = 0.02,\ \delta_h = 0.08,\ \pi = 0.3,\ m = 0.8.$$

There are two equilibrium points:

$$(k_1, H_1) = (17.858, 2.226),\ (k_2, H_2) = (10.401, 19.509).$$

The one with lower levels of human capital and per capita wealth is stable; the other is unstable. This conclusion is not dependent on specified parameter values. It is held in general conditions. This implies that in an economy with endogenous physical capital and endogenous human capital, the poverty trap is stably attractive, but prosperity is not easily sustainable. Figure 4.3 displays the vector field and the steady states of the dynamic system. As shown in Fig. 4.3, an economy with a low level of human capital, even if it was initially rich, tends to converge into the poverty trap. An economy with a high level of human capital, even if it was initially poor, tends to experience sustained growth. The nonlinear dynamic system has path-dependent features. Here, it can be seen the significance of cultural values for education. Japanese, Korean, and Chinese-dominated economies like Singapore and Taiwan could have sustained economic growth irrespective of their initial poor conditions in the 1960s, mainly because of their cultural values on education and the economic opportunities offered (guaranteed) by the America-dominated West. High saving rates did not play a crucial role in the determination of economic structural changes, even though many economists have argued for the importance of savings in those economic miracles. In the 1950s, no one could have foreseen the rapid economic development of East Asia, because few economists recognized the significance of education in economic development and fewer knew the validity of rationalism in those Confucian economies. In the mainstreams of economic development published in the 1960s and 1970s, capital accumulation is the main engine of economic development. Economists fail to properly interpret the economic evolution of these regions because they do not properly examine the cultural values of education in these regions. Under the great leadership of Chairman Mao, the mainland first eliminated the rich class and then eliminated the elite class.

I choose three initial states:

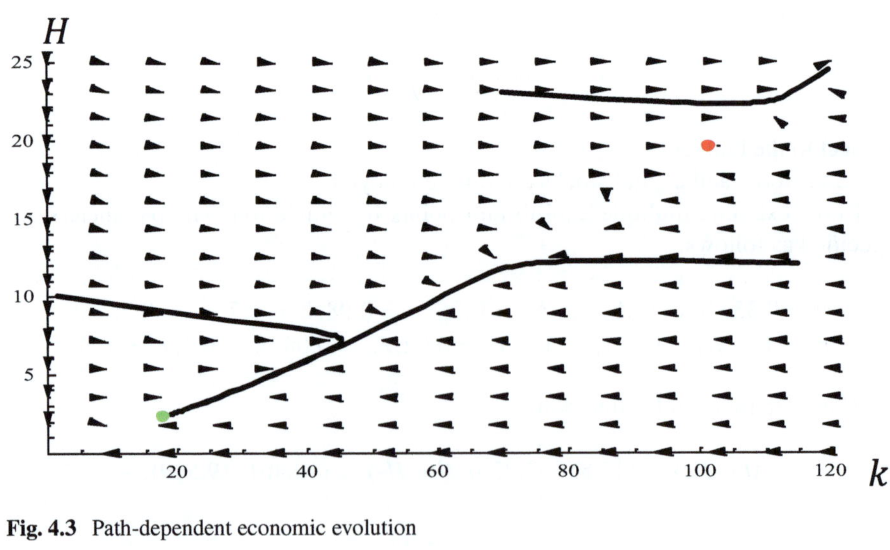

Fig. 4.3 Path-dependent economic evolution

$$(k_0, H_0) = (1.5, 10), (k_0, H_0) = (115, 12), (k_0, H_0) = (70, 23) \qquad (4.1)$$

The paths with $(k_0, H_0) = (1.5, 10)$ and $(k_0, H_0) = (115, 12)$ converge to the low levels of human capital and per capita wealth. It is interesting to note that the path with $(k_0, H_0) = (115, 12)$ starts with a high level of wealth. But its level of human capital has not improved over time. As decreasing returns dominate this path, its prosperity does not last long. The path with $(k_0, H_0) = (70, 23)$ will grow infinitely because the increasing return to scale dominates the economic evolution. This kind of infinite growth will not happen as the model neglects many other significant factors such as endogenous population change, negative externalities such as pollution, limitations of natural resources, and trade barriers. For instance, Taiwan may lose growth potential owing to its declining population and perhaps increasing trade barriers. Mainland China is also faced with aging and declining birthrates, an enlarged official population who enjoy secured welfare in a lifetime), and political uncertainties (with a bizarre ideology no properly educated man can logically understand).

As far as qualitative features of economic development are concerned, Fig. 4.4 provides some insights into the difference in the economic development in Mainland China and Taiwan during the period of 1950–1980. The two regions started their economic development with similar economic conditions but different average educational levels. Before the economic reform in 1978 started in Mainland China, the living conditions and educational achievements in the two Chinese societies had been greatly enlarged. It is only in recent years that mainland China has begun to explore opportunities for economic development. Structurally, mainland China's political-economic system had devaluated modern (Western) education so that no sector in the society could have explored the potential benefits of increasing returns offered by the Western civilization. Both cultural values and political systems matter in my model.

I now conduct a comparative dynamic analysis. Let us consider the case that the expenditure on education is reduced from 8% of the GDP to 7%: $\tau : 0.08 \Rightarrow 0.07$. Figure 4.4 shows the simulation results—the points with larger sizes are the new steady states and the other two points with smaller sizes are the old steady states. The two steady states are shifted as follows:

$$(k_1, H_1):(17.858, 2.226) \Rightarrow (14.758, 1.718),$$
$$(k_2, H_2):(101.401, 19.509) \Rightarrow (230.783, 53.417).$$

I see that the new stable steady state has lower levels of human capital and per capita wealth, but the new unstable steady states have much higher levels of k and H. It seems promising with the new education policy because the new, unstable steady state of higher k and H is much better than the old unstable one. Nevertheless, the economy with the discouraging policy has more chances to the traditional trap than to the economic miracle. For instance, if I start from (5.1.1), all the paths with these initial conditions end up in a poverty trap. The path with $(k_0, H_0) = (70, 23)$ exhibits the economic miracle. This example shows that the discouraging policy

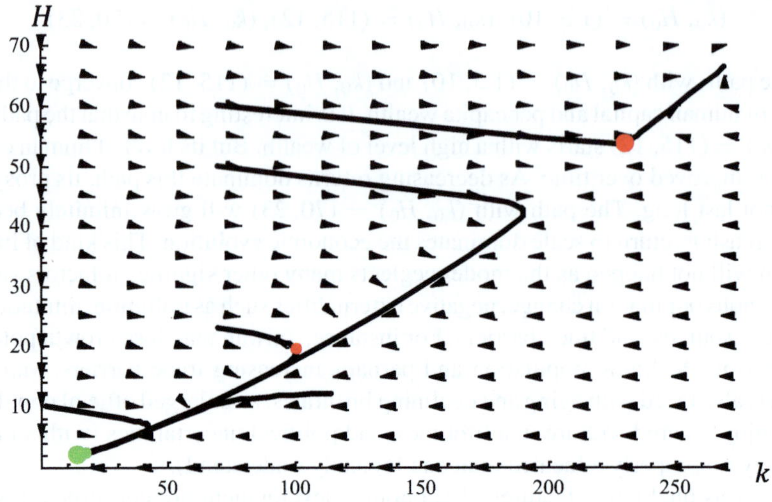

Fig. 4.4 The path-dependent development as education is discouraged

deprives society of development opportunities. The "chance" for development is lost due to the new policy that occurred in New China in 1949 when China's economy was controlled by the people with the traditional Chinese farmer mindset. As shown in Fig. 4.1.2, for the economy to experience sustained growth, the economy must have a much higher initial level of human capital than in the case of $\tau = 0.08$. Hence, if society reduces its investment in education, it will have much fewer opportunities to experience sustained economic growth, even though heavy investment in education will not guarantee the sustainable development of the nonlinear system in certainty.

I enhance the propensity to save. The propensity to save rises from 0.7 to 0.73, that is, $\lambda : 0.7 \Rightarrow 0.73$. Figure 4.5 displays how the system changes its paths. The points with larger sizes are the new steady states and the other two points with smaller sizes are the old steady states. The two steady states are changed as follows:

$$(k_1, H_1):(17.858, 2.226) \Rightarrow (14.758, 1.718),$$
$$(k_2, H_2):(101.401, 19.509) \Rightarrow (230.783, 53.417).$$

Figure 4.5 plots the effects of change in the propensity to save the dynamics of the system. The new stable steady state has higher levels of human capital and per capita wealth; but the new unstable steady state has lower levels of k and H. Without the Keynesian assumptions, I show that a higher propensity to save may not encourage economic growth.

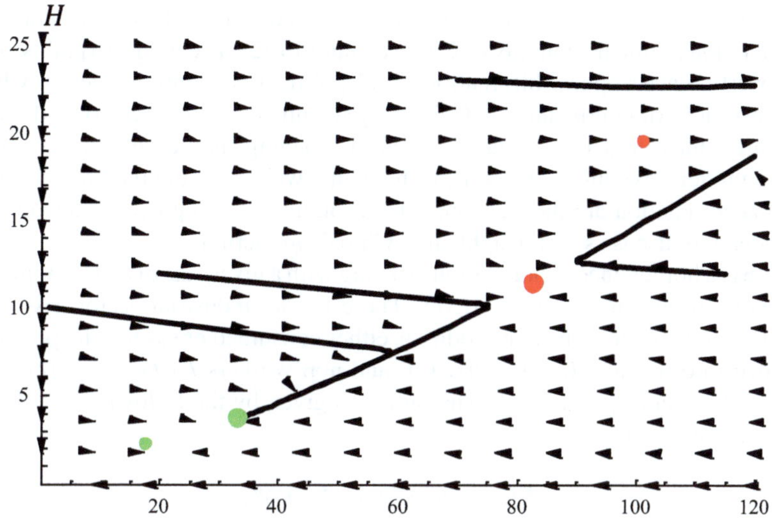

Fig. 4.5 An increase in the propensity to save

4.6 Catastrophe Theory and China's Economic Reform

B. R. Ambedkar (1891–1956) observes a trend in history: "History shows that where ethics and economics come in conflict, victory is always with economics. Vested interests have never been known to have willingly divested themselves unless there was sufficient force to compel them." Chaos theory teaches that causes and effects do not work in proportion. Great efforts may result in little fruits while little efforts may lead to great successes. India and China had similar economic conditions about 50 years ago, but today two economies have deviated hugely. The main cause for China's miracle is the introduction of a market mechanism—an idea even undergraduate economic students can give much better reasoning than the Chinese leaders who initiated the reform. The trivial idea in economics has caused structural changes in China—enabling one billion people to escape from the poverty trap.

China's economic miracle is affected by and affecting global economies in scale and scope unforeseen even two decades ago. The unexpected development was initiated by the economic reform in 1978. The main mechanism of structural change is a trivial idea in modern economics: market economy, which was theoretically proved (under certain conditions) and convincingly advocated. Before the economic reform, China had followed Karl Marx's socialism and had become poor, especially in comparison with not only developed economies but also Japan, Chinese overseas societies, and Korea. After China switched toward capitalism which Karl Marx admired for its efficiency, China has enhanced productivity, increased national wealth and GDP, and enlarged gaps of income and wealth, as well as social status as Marx, argued for the merits and demerits of capitalism.

About three decades ago, I proposed an economic growth model, which was devoted to the political economy of China (Zhang 1992). The model explains persistent poverty, catastrophes, bifurcations, and political-economic cycles in China's Westernization (modernization). Chinese government's attitudes toward market economies, measured in a variable called an opening, is the key mechanism for China's modern dynamics. Market mechanisms, wealth accumulation, and human capital accumulation are the key forces for economic development. I now describe the model and discuss some insights into China's modern history.

There is a homogeneous population growing with a given rate of n. The population changes according to: $\dot{L}(t) = nL(t)$. There is one industrial sector, like in the standard Solow model. The commodity is either consumed or saved. The population is divided into workers $L_1(t) = n_1 L(t)$, and non-workers $L_2(t) = n_2 L(t)$, where n_j is constant and $n_1 + n_2 = 1$. The output is given by the following neoclassical production function:

$$F(t) = K^{\alpha}(t)(z(t)L_1(t))^{\beta},$$

where F is the output, K the total capital stock, and z the human capital, $\alpha, \beta > 0$, and $\alpha + \beta = 1$. Let s denote the saving rate and introduce $k(t) \equiv K(t)/L_1(t)$. The capital accumulation is given by:

$$\dot{k} = sF(k, z) - (\delta + n)k, \tag{4.2}$$

where δ is the depreciation rate of physical capital. If z is not changeable, then the economy described so far is identical to the Solow growth model. There is a unique stable equilibrium.

In the case of China, human capital and knowledge have grown rapidly after the economic reform was initiated. Fast learning, as far as knowledge and technologies for industrialization are concerned, was not so much owing to creativity and learning from traditional Chinese knowledge stocks as from learning and imitating from foreign countries, including East Asian societies such as Japan, Singapore, Hong Kong, Taiwan, and Korea. It is those East Asian regions that made great contributions to China's economic growth in terms of physical investment, (low levels of) technologies, human capital accumulation, and cultural and institutional influences. We now consider changes in knowledge and human capital, $z(t)$. As mentioned before, a deterministic factor for China's economic success is its openness toward foreign societies. It will take one book to reveal how difficult and slow China had been in learning from the West (partly because China is a huge country, and the power group could live well by controlling a huge number of obedient people). I proposed the dynamics of the opening formed as follows:

$$\dot{p} = N\left[\epsilon p - \theta p^3 + q(k, z)\right], \tag{4.3}$$

where N is a positive speed adjustment parameter. The term $\epsilon p - \theta p^3$ represents the political forces that affect the opening of the nation. The linear term ϵp expresses the strength of reformers who support scientific and technological development and rationally learning from foreign countries, while the nonlinear term θp^3 represents the power of the conservative fraction who opposes foreign elements. The linear term implies that the more open China becomes, the greater the efforts of reformers are made. The nonlinear term means that the more open China becomes, the greater efforts of conservatives would be made in accelerating force after the nation has become very open. The relative strength of the conservatives to the reformers increases rapidly as China has become more open. A reason for this is that China has a long history almost isolated from the West and developed different cultural patterns with great achievements. Western rationality about man, society, and nature has had difficulties in penetrating those Chinese regions, which had little contact with foreign cultures. The term $q(k, z)$ includes factors that knowledge and living conditions of Chinese people affect the national openness. As living conditions are improved and knowledge has been diffused, the country tends to become more open to the outside world. The adjustment speed N is affected by many factors. For instance, during the decade of the Cultural Revolution, it is reasonable to consider N almost equal to zero. Nevertheless, during the early years of the economic reform, this number was large. There was little resistance against what was called foreign in mainland China. The situation is much different now.

There is a great increase in human capital and knowledge stock in China. There are sources of knowledge accumulation: learning by doing, learning from other cultures, and learning by consuming. Human capital accumulation follows:

$$\dot{z} = T\left[\frac{\mu p}{1+\sigma z} + rF(k, z) + vn_1(1-s)F(k, z) - \delta_1 z\right], \tag{4.4}$$

where T is adjustment speed, μ, σ, r, v, and δ_1 are positive parameters. The term $\mu p/(1+\sigma z)$ implies that a rise in the openness encourages human capital accumulation, but the impact is weaker if human capital is enhanced. The term $rF(k, z)$ simply means learning by doing. As workers produce more, they tend to have more skills. The term $vn_1(1-s)F(k, z)$ refers to learning by consuming (and education). If people consume more, experience more, and see more, they tend to have higher human capital. The term $\delta_1 z$ is the depreciation of human capital. I thus described the model which describes a nonlinear interdependence between wealth, openness, and human capital.

The period of isolation and poverty

Since New China was founded in 1949, China has been isolated from advanced economies. Under Mao's leadership, the wealthy class and elite were eliminated by socialist ideology. Mathematically, the economic growth is simply given by a single differential equation from (5.6.1), $\dot{k} = sF(k, z) - (\delta + n)k$. Since z is small, the system has a unique stable equilibrium. Since the equilibrium point was low and the population growth rate was high, China remained in a poverty trap until the other two equations began to be away from equilibrium with the economic reform.

Slow opening and catastrophe

I now examine how the catastrophe occurred in China's dynamics of opening. China has been under a single party's control. Whether the country is open and to what degree is determined either by national economic benefits or by formal laws. Everything is situationally dependent. The mindset of leadership and solidification of the party power is important in national policy. Given its population size, geography, and culture, China's political culture, different from Japan, Korea, and other Chinese regions (such as Hong Kong, Singapore, and Taiwan), had not been very adaptive. It is observed that for quite a long period, China's government emphasized Chinese characters and cultural special characters to resist Western influences. It is thus reasonable to assume, at least for explaining the history from 1840 to 1978, that the adaptive speed N is very small. It is so small that China's system achieves equilibrium in economic and knowledge systems. In long-term analysis, I can consider that the differential equations for human capital and physical are constantly at a stationary state for each given openness—the validity is mathematically discussed, for instance, in Zhang (1991). Rescaling time by $t^* =$ Nt, I rewrite China's political economy with the following systems:

$$0 \approx N\frac{dk}{dt^*} = sF(k, z) - (\delta + n)k,$$

$$\frac{dp}{dt^*} = \epsilon p - \theta p^3 + q(k, z),$$

$$0 \approx N\frac{dz}{dt^*} = T\left[\frac{\mu p}{1 + \sigma z} + rF(k, z) + vn_1(1 - s)F(k, z) - \delta_1 z\right].$$

As N is so small, I might consider that the first and third equations are always at equilibrium. It is thus, in an approximate sense, sufficient for us to be concerned only with the motion of openness:

$$\frac{dp}{dt^*} = \epsilon p - \theta p^3 + q(k, z).$$

This implies practically that it would be proper to examine China's modern history by focusing on its relations with the outside world because the system itself is "enslaved" to foreign influences during a certain period of its development. I now specify $q(k, z)$:

$$q(k, z) = b\big[c^* - n_1(1 - s)F(k, z)\big],$$

where c^* is the consumption level of the outside world (such as the United States or Japan). The term $q(k, z)$ measures the living conditions between foreign countries and China. This specified form tells that if China is poor, economic factors would

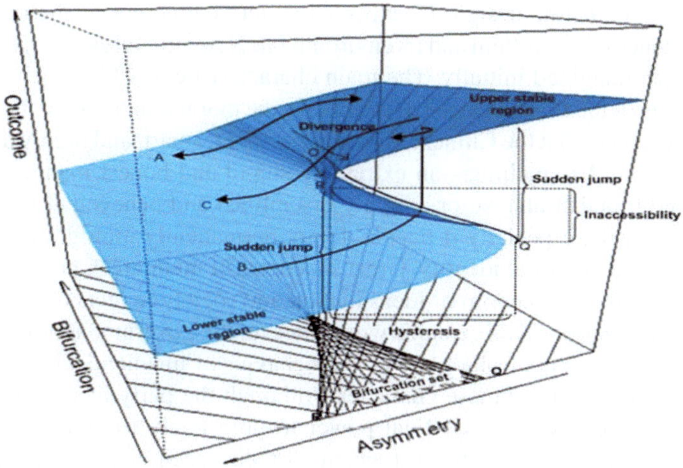

Fig. 4.6 China's cusp catastrophe in openness. *Source* https://www.ncbi.nlm.nih.gov/pmc/articles/PMC5664721

force the government to open the country. Under these preparations, it can be mathematically proved that cusp catastrophe occurs in China's political-economic system as plotted in Fig. 4.6, in which the vertical axis is the openness, and the parameter space is aggregated measurements of various economic, social, and cultural factors. When the catastrophe was occurring, China was very free with foreign ideas (and developed economies).

Three types of behavior are not examined in traditional comparative analysis. There is a sudden jump or catastrophe. In the case of this model, catastrophe brings about positive consequences to China's economic structure and development. There is a hysteresis which means that a reverse path to some point not being the same as the original. There is almost no way for China to return to the Cultural Revolution, even though some people are afraid of turning back. Cultural evolution is not continuous and reversible. There is a divergence which means that a small difference approach toward a cusp point leads the system to the upper or lower surface. Since the openness suddenly increased, human capital and physical capital dramatically increased. The economic miracle, which was brought about simply by opening the country with a market mechanism, has continued until today. There are hundreds of books and thousands of articles to describe this process. Complexity theory shows that people cannot accurately and convincingly describe the "mystery" of economic success. Fortunes and certain factors are mixed to determine the path of China's economic history in the last four decades. My model, which was perceived more than three decades ago, provides some structural insights into the process.

Process and order emerge not in a predictable or controllable manner. In the words of an American computer scientist, Alan Perils (1922–1990) says: "You can't communicate complexity, only an awareness of it." Emergence is one feature of self-organizing systems. Systems self-organize with paths, not according to an a priori planning or a deliberate design. Unexpected structure emerges as the result of the interdependence of the system and its environment. New structures emerge that could not have been imagined initially. The main characteristic of Chinese civilization is that the elite is rationally educated but the ruler is the moral monopoly of the society. It is rational to expect that the Chinese elite was obedient, smart, and highly disciplined and the ruler can be anything—an extremely stupid and honest man, a playboy, a cunning, and brutal strategist, or a highly disciplined and benevolent man. Culture does not die easily, especially if people do not permanently immigrate somewhere else to form a new cultural identity. Great successes of mainland China in almost all intellectual fields, but freedom of ideas in man and society reflects the continuation of Chinese civilization. This explains why whatever the government orders, Chinese society behaves with controlled order. The behavior of mainland China during the pandemic displays this Chinese character. Although the party has no well-defined goal or even sophisticated analytical power to objectively examine its long-term behavior, the "final objective toward which all their deceit is directed is to capture political power so that, using the power of the state and the market simultaneously, they may keep the common man in eternal subjection." (Henry, A. Wallace, 1888–1695). Since more than two thousand years ago, in China, it has not been objective law that limits the behavior of humans but the most powerful who takes the law into his hands. The law is secondary, the personal decision is above the law. This is similarly carried on in contemporary China, even though in some reformed form. Since it is obedience rather than freedom under the law, freedom has never been the main character in Chinese life. Servitude is rooted at the bottom of the heart.

Oscillations of China's openness with enhanced confidence and improved living conditions

China has been showing cyclical attitudes toward the West. There are political, social economic, and cultural reasons for changes in openness. In a large picture, China is no longer simply adaptive to and learning from the outside world. Hence, the model should be extended to include interactional relations with the outside world. Rather than entering too complicated interdependence, I limit myself to the original model to see why political oscillations have occurred. Adaptive speeds are not fixed. They are changeable. For instance, passion for learning may become slower. Corruption and inefficient application of knowledge are making Chinese youth lose passion and confidence in learning. This situation can be interpreted, in a nationally aggregated sense, that the adjustment speed of human capital is slow. That is, both N and T are small. Let $N = T$. Again re-scaling time by $t^* = \text{Nt}$, I rewrite China's political economy with the following systems:

$$0 \approx N\frac{dk}{dt^*} = sF(k, z) - (\delta + n)k,$$

$$\frac{dp}{dt^*} = \epsilon p - \theta p^3 + q(k, z),$$

$$\frac{dz}{dt^*} = \left[\frac{\mu p}{1+\sigma z} + rF(k,z) + vn_1(1-s)F(k,z) - \delta_1 z\right].$$

It is straightforward to show that the political-economic system is now given by:

$$\frac{dp}{dt^*} = \epsilon p - \theta p^3 + q(p,z),$$

$$\frac{dz}{dt^*} = q_1(p,z),$$

where I do not provide an explicit expression of q_1 because it is tedious. Under certain conditions, I (Zhang 1992) proved that there is a limit cycle $(p(t), z(t))$. There are periodic time-dependent relations between economic development, openness, and human capital in China's modern development. This analysis shows why China is so unpredictable but there are some regular patterns in a structural sense. A Chinese proverb says that dogs do not change their preference. This might also apply to some collective, i.e., the cultural, behavior of humans.

4.7 America's Greatness and Its Uncertain Futures

American philosopher Will Durant (1885–1981) points out a general pattern of civilization: "Civilization begins with order, grows with liberty, and dies with chaos." There are limited regular patterns of growth for a person, a region, a nation, and a race, even though we cannot be sure about accurate paths of evolution of a complex system. American poet H. W. Longfellow (1807–1882) reflects:

> Ships that pass in the night, and speak each other in passing, only a signal shown, and a distant voice in the darkness; So on the ocean of life, we pass and speak one another, only a look and a voice. Then darkness again and a silence.

A great thinker of Western civilization, known as the last universal genius, Gottfried Wilhelm Leibniz (1646–1716), who extensively studied things Chinese and perceived himself as a follower of Confucianism, predicted:

> It is in my view a unique disposition of fate which has placed the highest civilizations the human race has achieved as it were at the two extremities of our continent, that is in Europe and China, which adorns the opposite end of the earth as a kind of oriental Europe. And the highest providence is also at work in the fortunate circumstance that, while the nations which are most highly developed and at the same time the furthest separated reach out their arms to one another, everything that lies between them is gradually brought to a higher way of life. (Myers 1982: 160)

The center of European civilization shifted to America after WWII. The center of East Asian modernization was Japan for many years before Chinese and Korean regions sped up Westernization after WWII. Leibniz's view is still structurally correct. America is economically, militarily, and technologically, the greatest country that mankind has ever seen. Its value of freedom has deeply changed humanity. News about America is daily attracting attention on billion screens across the globe. It is simultaneously watched, studied, imitated, appreciated, targeted (for miscellaneous motivations), criticized, loved, mocked, and hated by someone somewhere ubiquitously on the Earth. English poet Oscar Wilde (1854–1900) describes America: "America is the only country that went from barbarism to decadence without civilization between." It does seem that contemporary China's speed is much faster—at least it should be, given the characters of modern civilizations—than America's. China has quickly been Westernizing but China is also picking up its cultural scums with an unbelievable speed. Traditional cultures tend to accumulate more scum than elements for a progressive future. Complexity theory shows that a complex system is full of chaos but there are structural attractors toward which it will move. In *America the Great and Its Self-Destruction* (Zhang 2021b), I reveal some mechanisms for America's sped up self-destructiveness. There is an increasingly deepening destructiveness in American economic, political, cultural, and social subsystems which are weakening America's relative position in the global division of labor and consumption. This section briefly shows the point.

Tocqueville (1835: 215) unearths: "I think that nations like men, in their youth almost always give indications of the main features of their destiny." He foretold that the United States would emerge as the leading naval power. He was certain that America was born to rule the seas long before that became a reality. Zinke (1868: 29) reveals how America became a great country:

> the wonderful progress of the United States, as well as the character of the people, are the results of natural selection; the more energetic, restless, and courageous men from all parts of Europe having emigrated during the last ten or twelve generations to that great country and having there succeeded best.... All other series of events - ... in the culture of mind in Greece, and ... in the empire of Rome - only appear to have purpose and value when viewed in connection with, or rather as subsidiary to . . . the great stream of Anglo-Saxon emigration to the west.

America has three core rules; capitalism, democracy, and individualism (+romanticism). As extensively discussed by Zhang (2021b), these values are no longer healthily synergetic in the American organ. A society has essential problems of creation, distribution, and consumption of wealth, power, and sex with self-interest, sympathy, and justice. These problems are interrelated in the long term. Separation of these problems can be temporarily effective for a young country like America. As society is more networked, a solution to one problem may cause more dysfunctions in other subsystems, just like the health problems of an aged sick man. The global game after WWII has been played with two fuzzy ideologies, communism, and capitalism. America was the main believer in capitalism. However, capitalism, like communism, includes self-destructive inclinations. According to the *yin–yang* vision of the world, socialism will move toward capitalism, and vice versa. Almost all main socialist economies have already moved toward capitalism either through quick collapses or gradual economic reforms. America cyclically moved toward socialism.

Its economic doctrine is Keynesianism, something a combination of communism and capitalism.

WWII facilitated America to establish a superpower. In late 1944, when WWII was ending, President Roosevelt (1944) was exulting: "At the end of this war this country will have the greatest material power of any Nation in the world. It will be a clean, shining America" (p. 405). The 35th president of the United States, John F. Kennedy (1917–1963) expressed American self-confidence: "We have the power to make this the best generation of mankind in the history of the world or make it the last." The United States is rich in natural resources. Its size and natural resources are important reasons for its rapid economic expansion since the nineteenth century. Protestant religions, the pioneer spirit of early settlers, and few restrictions on business ventures plus laissez-faire economic principles were basic determinants for economic advancements. Hughes and Cain (1998) identify a few principal elements that sustained the United States' economic expansion. The combination of available arable land and other resources, and an able population, gave America a history of unprecedented overall economic expansion. Those favorable conditions created a built-in optimism about the future. American freedom is manifested in several forms of liberty such as political, religious, moral, and economic freedom. Economic freedoms, such as freedom of competition, freedom of exchange between producers and consumers, buyers and sellers, free choice of one's vocation, free acquisition of rewards for one's efforts, and free accumulation of wealth, are the key elements in the liberty matrix. The United States appreciates freedom not only as a virtue, but also as a source of creativity, efficiency, and entrepreneurship. The economy has sustained economic progress. The sustained development is much owing to the synergetic effects of innovations, free competition, a market economy, immigration, a global environment, and monopolistic power. Americans were confident that they could build a good life with diligence. American optimism had been maintained by social mobility. It inherited no identifiable system of class barriers and no class of nobility. People born into poverty dreamed about getting rich through their efforts, and society encouraged them to make these dreams come true.

After WWII, America economically expanded. Many positive factors are combined to create the great epoch. Except for obvious reasons such as unpreceded scale and scope advances in applied sciences and technology, fast immigration is the unique determinant. Its population is now the third largest in the world. During and soon after WWII, armies of the best brains and well-known intellectuals were recruited by America from various parts of the world. In his Inaugural Address, on January 20, 1961, President Kennedy told the American people: "United, there is little we cannot do in a host of cooperative ventures. Divided, there is little we can do—for we dare not meet a powerful challenge at odds and split asunder. … And so, my fellow Americans: ask not what your country can do for you—ask what you can do for your country." World War II left a great opportunity for Americans to build a country with great universities, great projects, and great companies, to create global superstars in sports and arts, and to become a great collector of Nobel prizes. Voltaire (1694–1778) broaches why America could be united with white people who would have been enemies in Europe: "When it is a question of money, everyone is of the same religion." Tocqueville (1835) epitomizes the early American cultural traits:

> The passions that stir the Americans most deeply are commercial and not political ones, or rather they carry a trader's habits over into the business of politics. They like order, without which affairs do not prosper, and they set an especial value on the regularity of mores, which are the foundation of a sound business; they prefer the good sense which creates fortunes to the genius which often dissipates them; their minds, accustomed to definite calculations, are frightened by general ideas; and they hold practice in greater honor than history (p. 149).

American capitalism is primarily concerned with maximizing profit and one's utility. The democratic principles advocate for maximizing freedom, equality, and the public good. If capitalism widens the income inequality between people, then the democratic system may not tolerate enlarged income and wealth gaps. According to Okun (1975): "Such is the double standard of a capitalist democracy, professing and pursuing an egalitarian political and social system and simultaneously generating gaping disparities in economic well-being" (p. 1). The financial market has played a significant and special role in modern America. Greenspan (1996, 2013) uses the term "irrational exuberance" to expound on the behavior of stock market investors. In her popular *Makers and Takers*, Foroohar (2017) points out that Wall Street does not create jobs for the middle and working classes of America and that the financialization of America is threatening the American Dream. "The financialization of America," according to Foroohar (2017), "includes everything from the growth in size and scope of finance and financial activity in our economy to the rise of debt-fueled speculation over productive lending to the ascendancy of shareholder value as a model for corporate governance, to the proliferation of risky, selfish thinking in both our private and public sectors, to the increasing political power of financiers and the CEOs they enrich, to how a 'market knows best' ideology remains the status quo, even after it caused the worst financial crisis in seventy-five years." (pp. 5–6).

Household income and distribution are important variables for social welfare. In America, a household's income is the total income of every resident over the age of 15, including pretax wages and salaries, any pretax personal business, investment, or other recurring sources of income, and any kind of governmental entitlement such as unemployment insurance, social security, disability payments, or child support payments received. As plotted in Fig. 4.7, from 1968 to 2018, the share of the national total income of the highest earning 20% of American households has steadily increased. By 2018, households in the top 20% earners brought in 52% of all US income, while in 1968, that number was only 43%. The top 5% of households had their share, 23%, of all US income, while that number was only 16% in 1968.

America has now the best education system in the world. Most of the 100 top universities ranked are American. It caters to its citizens as well as global students with the best education. Figure 4.8 shows that household income and per capita income are positively and significantly related to educational attainment.

Figure 4.9 (Leiserson et al. 2019) plots the dynamics of wealth distribution from 1989 to 2016. The wealth distribution also verifies that the top 1%'s share has increased rapidly in recent decades.

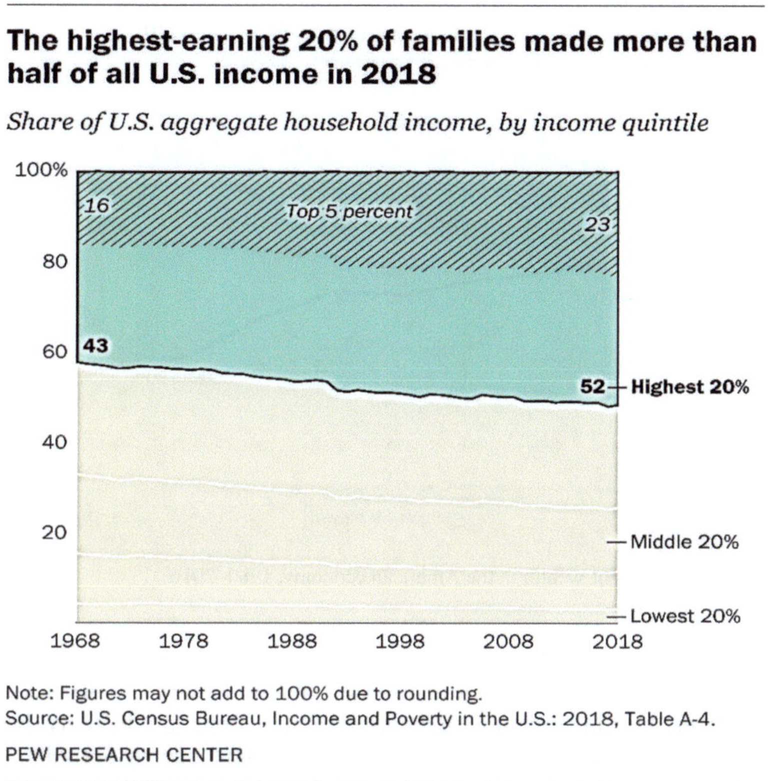

Fig. 4.7 The dynamics of American income distribution from 1968 to 2018. https://www.pewresearch.org/fact-tank/2020/02/07/6-facts-about-economic-inequality-in-the-u-s/

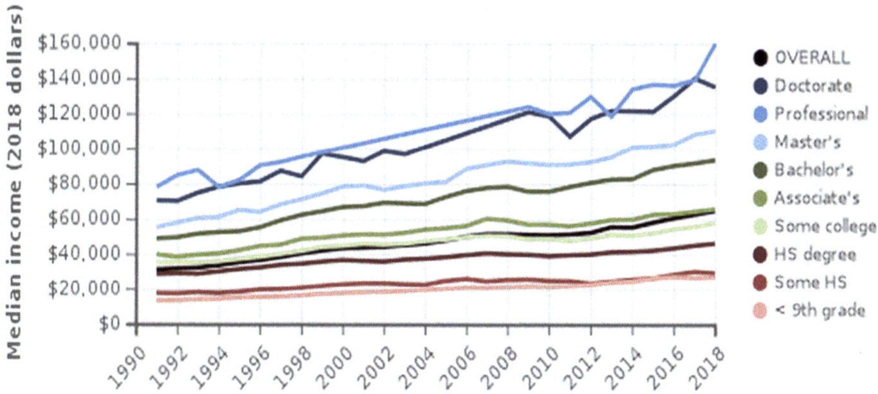

Fig. 4.8 Household income distribution due to education degree. *Source* https://en.wikipedia.org/wiki/Household_income_in_the_United_States

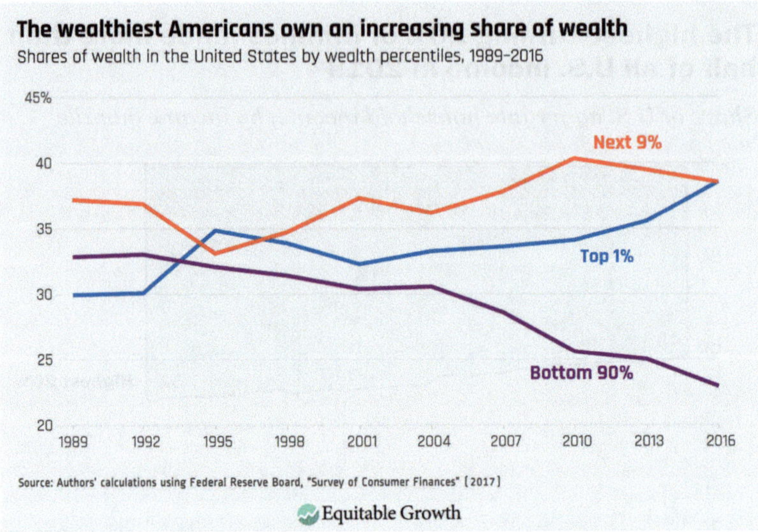

Fig. 4.9 The dynamics of wealth in the American economy, 1989–2016

Figure 4.10 plots growth rates (%) of productivity and worker's hourly compensation for America's economy. From the late 1940s to the early 1980s, the pay of typical workers rose in tandem with the increase in national productivity. But after the 1980s, America's national productivity has continued to increase rapidly, and the rise in speed in the wage has fallen far behind. This implies that American workers might have worked more productively, but their contributions to the economy are not fully shared by them but parts are accrued to those at the top and corporate profits.

"American culture is CEO obsessed. We celebrate the hard-charging heroes and mythologize the iconoclastic visionaries." (Marcus Buckingham 1966). Figure 4.11 plots the aggregated CEO-to-worker compensation ratio for the 350 largest publicly owned companies in the United States from 1965 to 2019. By 2019, the ratio reached

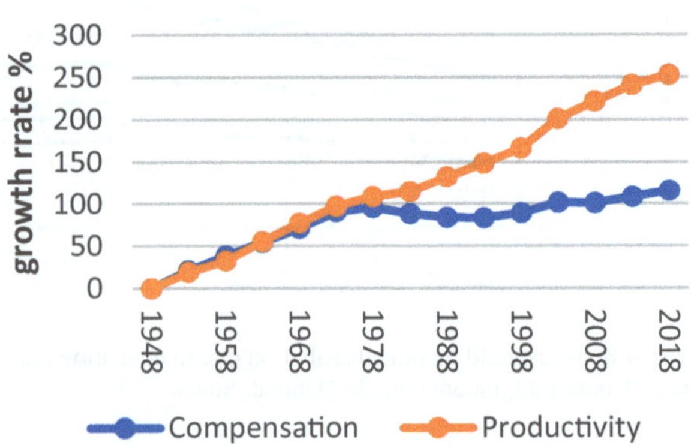

Fig. 4.10 Growth rates (%) of productivity and worker's hourly compensation. *Source* https://www.epi.org/productivity-pay-gap/

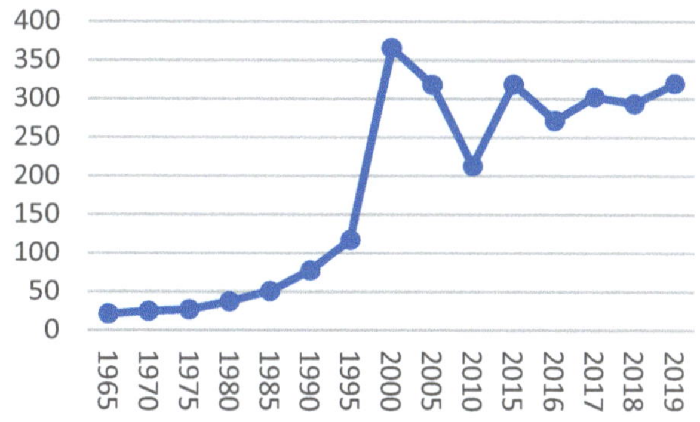

Fig. 4.11 The compensation ratio of CEO and worker, 1965–2019. *Source* https://www.statista.com/statistics/261463/ceo-to-worker-compensation-ratio-of-top-firms-in-the-us/

320. On average, CEOs received about 320 times the annual average salary of production and nonsupervisory workers in the key industry of their firm. American CEOs collect large paychecks, stock options, bonuses, luxury gifts, and companies' other unseen benefits. Some CEOs take huge incomes without any sense of duty when much of the income is due to the contributions made by those who are losing their jobs.

Bourdieu (1998: 27) observes:

> … worn out parents, exhausted young people, employers disappointed by the products of an education which they find ill-suited to their needs, are all the helpless victims of a mechanism which is nothing but the cumulative effect of their own strategies, engendered and amplified by the logic of competition of everyone against everyone.

Since the 1980s, relative poverty rates in America have been higher than those of other wealthy economies. Extreme poverty, which is defined in the United States as a household living on less than $2 per day before government benefits, reached 1.5 million households with 2.8 million children in 2011, doubling the level in 1996. As of 2015, 44% of children in America lived with low-income families, as shown in Fig. 4.12.

Household saving behavior is reflected in the personal saving rate. Figure 4.13 exhibits the dynamics of personal saving rate—the ratio saved by individuals or families to the disposable income—of the United States, 1960–2019. In recent years, American households save more out of their disposable income. Except for some determinants for this tendency, an important factor is often mentioned. As income and wealth distribution is enlarged, the poor have nothing to save, and the rich have too much to consume. For instance, in 2018, over 10% of American adults would be unable to cover $400 emergency expenses.

Innovations for improving productivity are encouraged in America's free markets. There were great American innovations, such as the Franklin stove, Whitney's cotton gin, Fulton's steamship, Edison's electric light, and the Bell telephone, which have shaped the world. American great achievements in science and technology are uncountable. America-made innovations in the areas of cars, aircraft, computers, and antibiotics that are globally used today. These technological advances are closely related to social change (e.g., Harrington 2008; Stein 2019). Cowan and Hersch

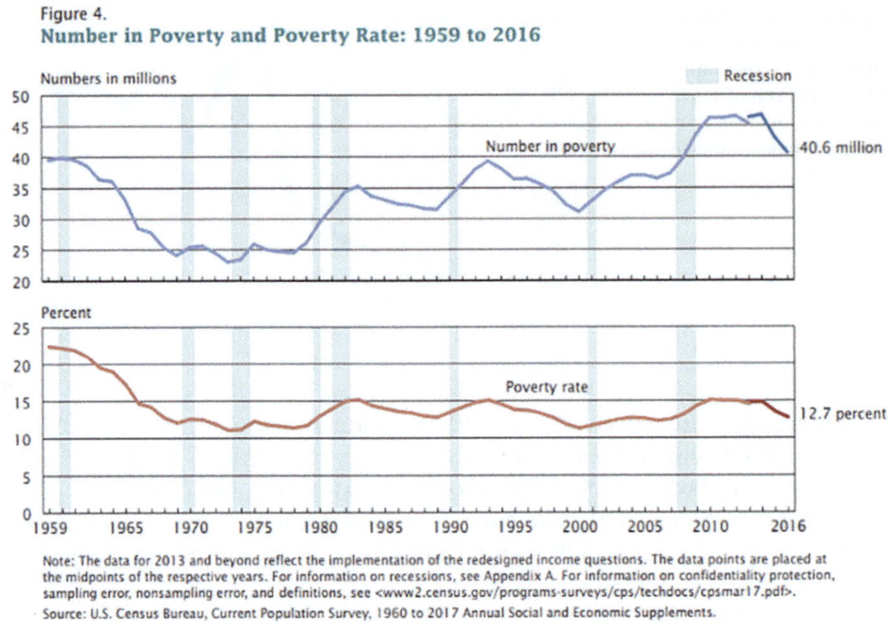

Fig. 4.12 Poverty and poverty rate of America, 1959–2016. *Source* https://en.wikipedia.org/wiki/Economy_of_the_United_States

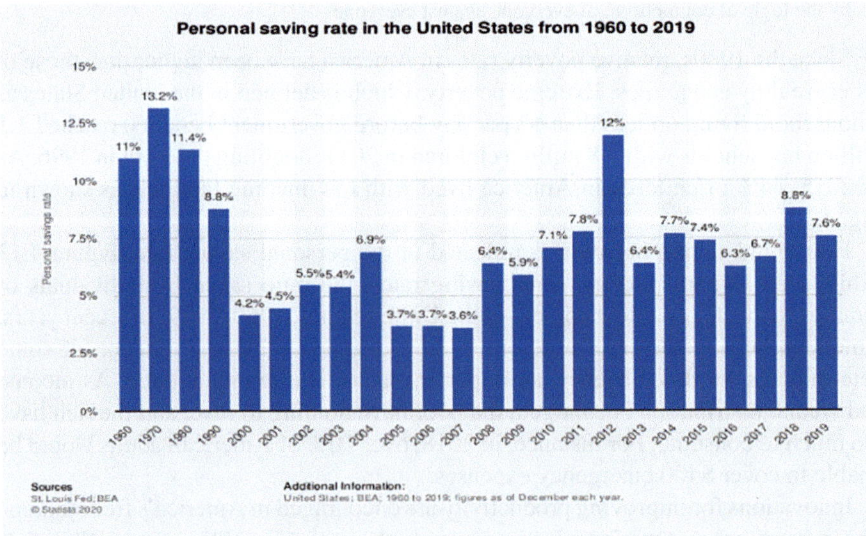

Fig. 4.13 Dynamics of personal saving behavior in America, 1960–2019. *Source* https://www.statista.com/statistics/246234/personal-savings-rate-in-the-united-states/

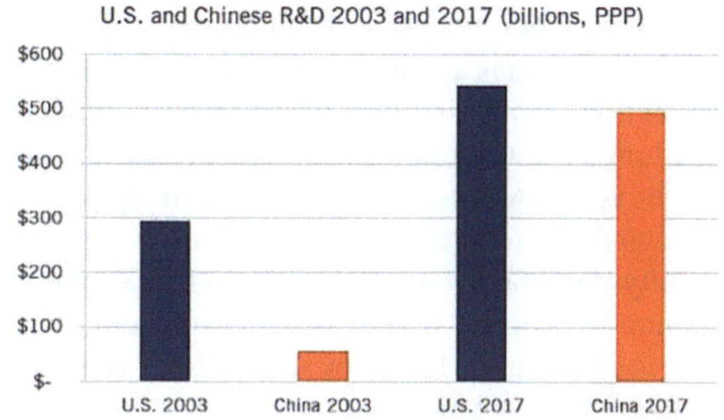

Fig. 4.14 China is catching up with America in total R&D expenditures. *Source* https://itif.org/publications/2019/08/12/federal-support-rd-continues-its-ignominious-slide

(2017) study the interdependence between technological change, American society, and American geography. America excelled in applied sciences and technologies. Figure 4.14 plots the expenditures of the two greatest economies on R&D. The figure explains why China is catching up both in firms-based technologies and nation-level projects in space and infrastructures. As the two countries' GDPs are approaching and expenditure structures are similar, the rivalry between the two should evolve severely in many fields, especially because China has many talented people (many of them received higher education in America) and workers are energetic and lowly (with regards to America) paid.

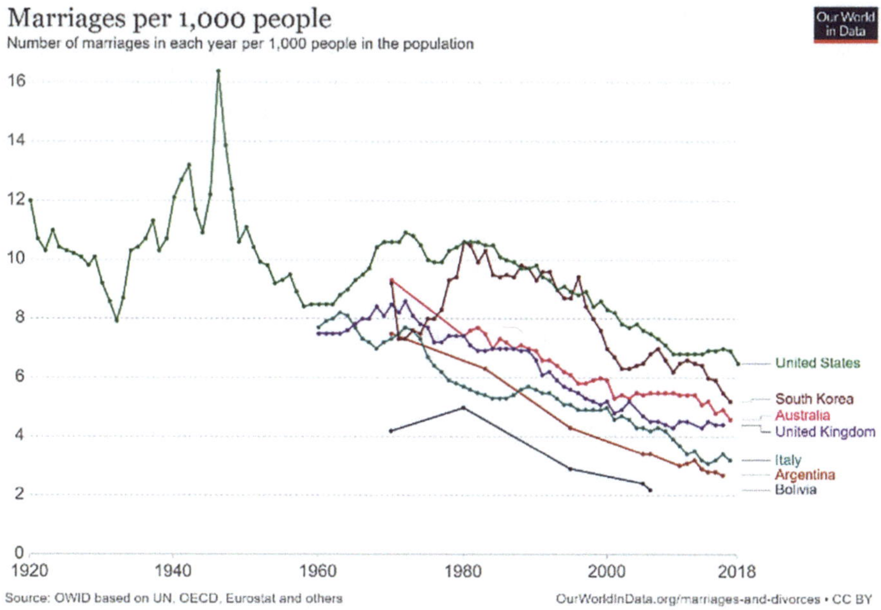

Fig. 4.15 Dynamics of marriage, selected countries

Deng Xiaoping (1904–1997) epitomizes the American political practice: “The United States brags about its political system, but the President says one thing during the election, something else when he takes office, something else at midterm and something else when he leaves.” The president could serve the country a maximum of eight years. A president might love to spend public money as much as possible to establish his reputation. America borrows from the world without even thinking carefully about how to pay for it. Soren Kierkegaard (1813–1855) describes behavior of common people: “People demand freedom of speech as a compensation for the freedom of thought which they seldom use.” Bernays (1928) penetrates the formation of public ideas in America:

> It may seem an exaggeration to say that the American public gets most of its ideas in this wholesale fashion. The mechanism by which ideas are disseminated on a large scale is propaganda, in the broad sense of an organized effort to spread a particular belief or doctrine.

There are two major political parties, the Democratic Party (1830) and the Republican Party (1854), in America. Niccolo Machiavelli (1469–1527) tells honestly the truth of the human race: “Politics has no relation to morals” Mark Twain sees the continuation of American politics: “We have the best government that money can buy.” American political game has been played largely between the two parties. Game theory and the psychology of masses mathematically verify that no party should have a solid, rational principle to abide by. One might consider a party as a firm in a market economy. Its strategies are political ideas and promises about benefits and its award (power) is the number of votes. Mass’ emotions are in flow without a fixed principle. Collective emotions and weak rationality (and shallow knowledge) determine who wins the game. Alexis de Tocqueville observes the American character: “I know of no country in which there is so little independence of mind and real freedom of discussion as in America.” One can still feel the continuation of American culture in TV debates. It makes people, who are not yet accustomed to American culture, feel that American people debate in a way that the maximum number of words can be regurgitated from their mouths.

A great mind of the West, Ludwig Wittgenstein (1889–1951), says: “I don’t know why we are here, but I’m pretty sure that it is not in order to enjoy ourselves.” Human society is a biosystem. Its continuation is based on the production and reproduction of people. A race that makes a great contribution to mankind may disappear if it fails to reproduce itself. Among the three key objectively measurable variables—wealth, power, and sex. Sex and its associated products, such as family, children born outside marriage, and adultery, are essential for reproduction. Jordan Peterson (1962) argues:

> We’re so immaturely cynical as a culture. We’re not wise enough to look at an institution like marriage and to really things about what it means and what it signifies. It signifies a place where people can tie the ropes of their lives together so that they’re stronger. It signifies a place where people can tell the truth to one another.

In modern times, people choose to live together without marriage, marry late, and divorce. As demonstrated in Fig. 4.15 (Ortiz-Ospina and Roser 2020), in 1920, marriages per 1,000 people in the United States were twice as today. The long decline started in the 1970s and is continuing.

Family time refers to the time that family members share. Children are proximally influenced by parenting through contact and interactions with the parent. Parents provide an environment and conditions for growth, such as shelter, neighborhood, food, and educational opportunities. Time and allocation of time for valuable for children's development as well as the relationship development of adults. Family time is essential for organizing collective family activities. Collective family gatherings such as mealtimes, are commonly observed in traditional societies. In association with the Industrial Revolution, families, for instance, in Britain, changed their daily lives as work patterns outside the home shifted. Except for holidays, family members had daily routines: going to work or school, having the family dinner, doing homework, and going to bed. The structure of daily activities and routines in the home is influenced by social and economic forces outside the home. It is observed in contemporary America that people in poverty are faced with escalating family chaos whereas well-educated and prosperous families have stable living conditions for their children. There are three common desires in human society—the desire to own wealth, the desire to be powerful, and the desire to have sex and establish and extend sex-based trustful networking. The core values of production and consumption of these goods and services in the United States are correspondingly democracy, free market, and individualism (human rights, gender equality, liberalism, romanticism …). The desire for sex seems to be relatively easily solved by freedom of choice and a free market mechanism in a modern society with huge wealth and modern technology. Even the Chinese Communist Party "allows" the gender market to self-organize with freedom between heterogeneous genders (not yet modernized to the same gender and marriage between the same gender) with legally free and low costs of divorce.

Family chaos, which might result from sensory overload, physical crowding, and routine family life, has different effects on members of the family (Ward 1995; Fiese and Winter 2010). Necessary routine family work is largely reduced owing to technological changes. Scale and scope economies work in different ways from traditional societies as consequences of economic development and technological change. How families live together with healthy and happy lifestyles for everyone varies over time and across space. Married couple relationships, parent–children's relationships, and the human capital of each member are essential to the health and well-being of each member. Family chaos, family crisis, marriage between the same gender, divorce, children brought up by a single parent, and the like, are popular terms to describe modern family chaos. Family wealth, power, and sex are closely interrelated and essential for any society. The family tied by blood is composed of several people. In a modern society, for instance, America, where family is not "managed" according to traditional customs and religions, the members of the family live together like playing a game—each individual makes a decision according to his own rational as well as emotional calculations. Cooperation, competition, and negligence of the existence of other members display a great variation in American societies owing to their individualism, coexistence, and multiple cultures. Life in metropolitan areas often lacks structure and is chaotic in everyday activities. The family itself faces external environments such as neighborhood, work environment, schools, taxation, and other institutional rules. Family chaos reveals in the disruption

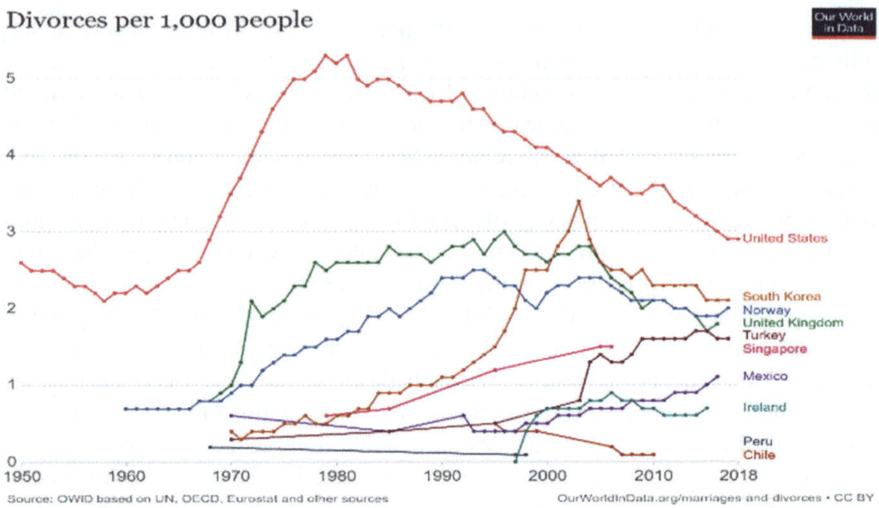

Fig. 4.16 Dynamics of divorce, selected countries

of daily activities and the subsequent interpretation of the event as a whole. Fiese and Winter (2010) discuss three diverse but interrelated levels of family chaos: the construction of family time, the frequency and disruption of family activities, and the meaning created out of disrupted or irregular activities. The high divorce rate follows American democracy and individualism, especially when America goes through fast economic progress and rapidly enlarging income gaps. The American value of independence, the pursuit of happiness, geographical mobility, and prosperity should have a natural equilibrium divorce rate in the gender market, as demonstrated in Fig. 4.16 (Ortiz-Ospina and Roser 2020).

Romanticism under the protection of the law, which has not much to do with human nature and freedom, is a modern innovation. All rational and prosperous economies with romanticism are either importing a large portion of immigrants or are faced with serious dilemmas in association with the falling population. Immigrants might contribute to American society far less than they cost because simple jobs can be replaced by robots and lifetime costs, associated with poverty, education, medical care, and other social welfare are rapidly increasing in developed economies. It is not so relevant to assume that highly talented and diligent foreigners would choose America as their homeland like in the last century. It is not an easy matter to deal away with drugs, especially for a country with long prosperity and freedom like America. Drug equips one to lead a fast, frivolous, and dissipated life. It has a great power over men and the attractivity naturally lasts longer than carefully faked sexuality. It is well-known that opium made a great contribution to the history of China's humiliation from the 1840s till the 1910s.

Criminal is an integrated part of human society. In *Meditation*, Marcus Aurelius says: "To expect bad men not to do wrong is madness." Emerson (1803–1882) describes American reality: "Good men must not obey the laws too well." This is

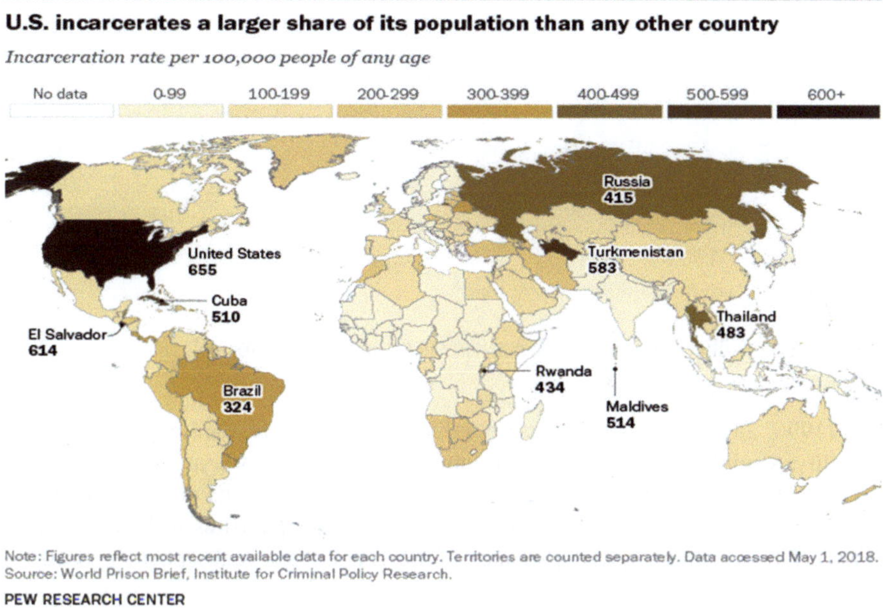

Fig. 4.17 Incarceration rates of the World, May 1, 2018. *Source* https://www.pewresearch.org/fact-tank

a practice-accepted definition of the good and practically active American man, as distinguished in the behavior of America-made heroes such as, Benjamin Franklin and Steve Jobs. In recent years nearly seven of every thousand people in the United States are in a federal or state prison or local jail. The number rises to 11 of every thousand working-age adults. America has less than 5% of the global population but its share of the world's incarcerated people is 20%. Like its Nobel prizes, America has a much higher rate than the average rate in the rest of the world. America's relative position in the world is illustrated in Fig. 4.17. Contributions of the military sector, drugs, prostitutes, crimes, and private guns to the US economy are so huge that the government would hardly do anything with its huge income resources.

America's economic superpower is now tested by many countries. In 2018, its nominal GDP and PPP were $21.44 trillion; the corresponding variables for China are respectively $14.14 trillion and $27.31 trillion and for Japan, $5.15 trillion and 5.75 trillion. China's PPP has exceeded America's for some years already. In 2018, America accounted for 15.2% of the global gross domestic after adjusting for PPP. Figure 4.18 plots the dynamics of the share of America's economy in the global economy during the period 1960–2014. In half a century, the share would fall from 40% to nearly 20%. The process of declination is still going on.

America had been predominantly Protestant in the early stage of the construction of the American heart. Many of its business leaders have attributed their economic success to the Protestant ethic in Weber's terms (McClosky and Zaller 1984: 104).

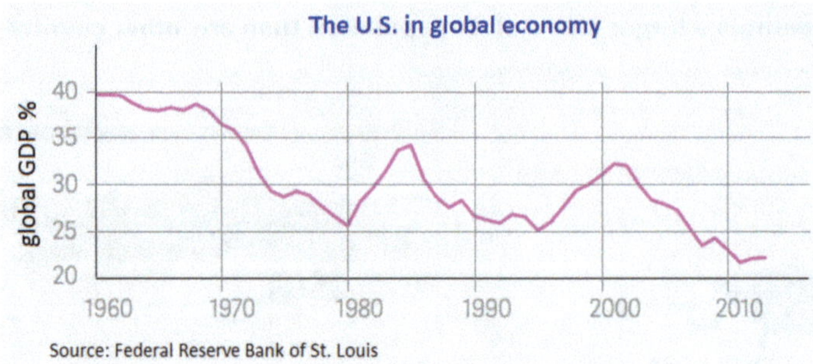

Fig. 4.18 The falling share of the U.S. economy in the global economy

Bell (1996: 289) sees: "The Protestant ethic as a way of life, then, was one of piety, frugality, discipline, prudence, the strenuous devotion to work, and delayed gratification." Human character changes over time. New generations do not follow their parents' habits and customs in modern societies. Like parasites in human bodies, a society feeds many "social parasites." They also might make positive contributions to socioeconomic dynamics. Lawyers, medical doctors, and financers are very expensive for society to keep. A great lawyer might be a great liar and his highly paid achievements lower the morality of society. A successful medical doctor may be highly paid and not available for real workers but employed by the aged rich (due to heritage) without any contribution to the national future. High talent is wasted in multiple ways in serving social "parasites." Many official administrators, lawyers, and medical doctors may make little if not negative, contributions to social and economic advancement in the long term. Many harmful parasites enjoy prosperity and evolve to use up social resources. A society that is strongly influenced, manipulated, or even controlled by a small number of social parasites will be ruined, soon or late. William Banting (1797–1878) notices: "Yet the evil still increased, and, like the parasite of barnacles on a ship, if it did not destroy the structure, it obstructed its fair, comfortable progress in the path of life."

Lincoln's "Of the people, by the People, for the People" has encouraged people not only in America but also people outside America, to build their countries more democratic and more equal. Contemporary America is further away from the ideal. The American government is by the rich and mainly for the rich. The influential Nobel Prize winner in economics Stiglitz (2013) illustrates the obvious:

> The simple story of America is this: the rich are getting richer, the richest of the rich are getting still richer, the poor are becoming poorer and more numerous, and the middle class is being hollowed out. The incomes of the middle class are stagnating or falling, and the difference between them and the truly rich is increasing.

Emerson (1803–1882) represents the honesty of early American thinkers: "Truth is beautiful, without doubt; but so are lies." In *The Elephant in the Brain*, Simler and Hanson (2020) depict a common human trait:

> Human beings are primates, and primates are political animals. Our brains, therefore, are designed not just to hunt and gather but also to help us get ahead socially, often via deception and self-deception. But while we may be self-interested schemes, we benefit by pretending otherwise. The less we know about our own ugly motives, the better – and thus, we don't like to talk, or even think, about the extent of our selfishness. This is 'the elephant in the brain.'

American civilization is weakening its greatness due to its greatness. For the Chinese *yin–yang* world vision, self-destruction is a natural process of creativity and construction. Love lasts short, while hate can be easily protracted. Sympathy for poverty needs the existence of poverty. Liberty versus slavery, democracy versus hypocrisy, individualism versus cruelty (to the outsider), and the like, are the two sides of the same coin. If the positive side is strong, the negative side is seemingly weak for a while but will rebound so that the organic system is dynamically balancing. Fulton J, Sheen (1895–1979) holds:

> The Tragedy of our time is that those who still believe in honesty lack fire and conviction, while those who believe in dishonesty are full of passionate conviction.

The phenomenon is perhaps not necessarily "our time" as any country, which has a period of high civilization experienced. It is a natural equilibrium in high civilization because one who understands (man and society) tends to act as little as possible and one who understands a little will passionately act. No culture with a history of higher civilizations could have avoided the status of the civilized equilibrium like no aged man can avoid slow but steady health degeneration if not sudden death.

4.8 Complexity in Politics of Knowledge-Based Societies

Complexity theory has turned out to be an effective way of examining webs of political intrigue (e.g., Cairney 2012; McGee and Jones 2019). National policy, such as taxation and wars against foreign countries, usually involves almost all members of the nation. Political consequences are often beyond what the decision-maker has expected because the system is complex, and the decision is related to future events. For instance, a free market mechanism with proper government planning may be successful in China but become catastrophic in India. Grand ideologies or national projects do not follow what they are supposed to achieve. Socialism, which is supposed to build a happy and equal society led to equal poverty and low productivity. Capitalism has failed to deliver welfare to the majority of the population and is enabling the rich to become richer and the poor to become poorer.

Modern society creates, diffuses, and applies knowledge at a speed unprecedented in human history. Expansion and spread of knowledge have enriched political gaming at all levels of society. Knowledge is power for enriching oneself and for one to utilize the environment. Children learn skills and pursue knowledge with high curiosity. Sydney states: "If you want to be truly successful invest yourself to get the knowledge you need to find your unique factor. When you find it and focus on it and preserve your success will blossom." To know is not only to know what can be done and should be done as perfectly as possible, but also to know what should not bother about or try your best to avoid. Knowledge is the power to effectively employ or be employed by external factors.

Man is the same at birth and has nothing in this world. The final equilibrium point is also identical. It is the process of life that characterizes and separates men. In a free fair social and economic environment, one's social status, given one's born nature and character, is determined not so much by other people and the local environment, but by how one spends it—educational, professional, and leisure time, by what kind of people one chooses to be associated, by what will or goal(s) one builds for oneself, what knowledge and skills one accumulates over time, by what ability one can effectively apply one's knowledge and skill, by how one cares about the health and gives one's health to the will, by how free and creative one's mindset. Life path is affected by fortunes or bad luck which are beyond one's control, but one can avoid many misfortunes and grasp one's luck if one has a prepared mindset. Theodore Roosevelt describes knowledge and virtue: "A man who has never gone to school may steal from a freight car; but if he has a university education, he may steal the whole railroad."

People communicate, cooperate, and compete in various ways. It is commonly believed that knowledge is a powerful tool to change the world. Nowadays information is more powerful than knowledge to influence occurrences of personal, social, as well as global affairs. Judgments and decisions are made not on in-depth knowledge, but on conspiracy theories, manipulated information, and fake news. Nicholas M. Butler observes: "America is the best half-educated country in the world." This is the reason of all the reasons that America has been able to maintain its superpower. Any judgment is measured by the value system for measurement. Men comprehend man and society through reading books and observing what they are. American economist Thomas Sowell observes American education: "Education is not merely neglected in many of our schools today but is replaced to a great extent by ideological indoctrination." In interactions, Elias (1992) points out that people need:

> … a symbol of a relationship that a human group of beings biologically endowed with the capacity for memory and synthesis, establishes between two or more continua of changes, one of which is used by it as a frame of reference or standard of measurement for the other or others (p. 46).

When the size of society becomes large, it needs symbols that can hardly be identified with certainty by individual members so that the leader can control the masses by interpreting the symbol according to changing conditions. From a long-term perspective, no symbol is permanent. Nevertheless, the importance of such a symbol is emphasized by Elias (1992: 9):

> Temporal norms would seem to play the imminently social role of guaranteeing the organization of work, the systematic satisfaction of reciprocal expectations in people's behavior towards each other, at the same time as they express evaluations and moral positions in the face of the fundamental experience of change and the awareness of death.

Mankind has suffered much in the last century in selecting "symbol systems" without any lasting meaning. Globalization and localization in symbol systems are continued without end.

Nietzsche reveals: "Not when truth is dirty, but when it is shallow, that the lover of knowledge is reluctant to step into its waters." Knowledge—whatever motivation one has for accumulating it—is through oneself. Politics, like love and stock markets, is full of chaos. Power, sex, and wealth are three key observable and measurable variables for understanding the evolution of human society. Modern democratic societies spend much social energy and money on electing leaders. As society becomes more complicated, governments must address more problems. Even before they are recognized and solved by governments, unsolved problems interact and new problems emerge. Democratic leaders, if examined in detail by the path of their past speeches and opinions, make mediocre and even nonsense talks most time in most circumstances when the society is unstable and faced with many problems. They appear effective and clever when they are not much needed by the society. There is no certain correspondence between the national needs and the elected human quality. Those whose human capital is perhaps lower than a smart graduate student in social sciences in a globally reputed university get final victories by manipulating the masses' emotions with unachievable promises. Few democratic leaders are respected, at least in respectable countries, by those who elected them after being in power only a few years. In many situations, the political game is to deceive the deceiver. The political game is run by man's nature of deception and manipulation. Fictions, national faked heroes, unexamined principles, blind faiths, and the like are foundations of highly civilized nations. The lie is accepted as faith or even truth if it is repeated with confidence over time.

Saperstein (1999) describes the necessity and limitations of modeling in political science as follows:

> Dynamical modeling is an important component of verbal pollical science. 'Modelling' refers to the creation of a representation of the world of interest - in your mind, on paper, or in the laboratory. … [Y]ou use a representation, a partial world that you hope contains the aspects of that world important for the behavior that you wish to understand. … Again, modeling is a necessary characteristic of conventional science, though usually informally and implicitly.

The political game is difficult to model because politicians often design rules of games that they will play, judge, and benefit from. Popular political theories are often built on the assumption that politicians would decide for the benefit of the society. Wrong assumptions do not lead to correct explanations of the phenomenon. Wrong applications of the assumptions of a theory may lead to miserable results. For instance, China had wrongly applied Kar Marx's ideas about practicing socialism and the nation had been miserably punished. It is through applying Marx's ideas about the positive forces of capitalism that China has succeeded in economic development.

In a knowledge-based society, education provides a basic input to the performance of politics. Politicians manipulate society through the elite class. Education and education degrees play a vital role in social and economic positions in modern economies. Different from traditional education which emphasizes faith and virtue, modern Westernized education puts more weight on knowledge and skills for becoming professionals. About university education, George B. Shaw comments: "A fool's brain digests philosophy into folly, science into superstition, and art into pedantry." It is crucial for a modern society that people get properly trained for building mindsets for the future. If the elite is good at nothing but limited professional knowledge and skill, politicians could lead the human society like an ant society which may self-organize into catastrophes. Education is not a perfect equalizer of the conditions of men even in the long term. A university should not be a place where students spend four years only accumulating knowledge. It is a great opportunity for one to enter another stage of life. It is a community and cultural environment in which students, teachers, alumni, administrators, supporters, locals, and the like are mutually connected.

Sigmund Freud interprets: "The interpretation of dreams is the royal road to a knowledge of the unconscious activities of the mind." Spread formal education can lead to unexpected consequences and catastrophes. With regards to Indian education, for instance, Sugata Mitra observes: "The Indian education system, like the Indian bureaucratic system, is Victorian and still in the 19th century. Our schools are still designed to produce clerks for an empire that does not exist anymore." Experts in narrow fields without self-cultivation and moral education turn out menaces to human societies. Blaise Pascal describes: "Knowledge of physical science will not console me for ignorance of morality in a time of affliction, but knowledge of morality will always console me for ignorance of physical science." One might characterize modern society by that there is a lot of sexuality but not much love, there is plenty of knowledge but not much wisdom, there are many friends but not much trustfulness, there are many kind and polite words but not much heart, there are full of elected politicians but not much principle, there is a flood of information but not much trustful message.

Public management is an important part of whether political decisions can be effectively carried out. Complexity theory studies public management (EL-Ghalayini 2017). It deals with the adaption of environmental change and the complexities of interactions between different parts of public organizational systems. Public management is treated as a process of interdependence between many agents. The main streams of traditional public management theories are dominated by stability, no structural change, and a lack of adapting environmental variety. Political organizations can be considered as a complex adaptive system or a part of the national complex system. Complexity theory provides new insights into the complexity of politics across countries over time. It gives new opportunities to understand the evolution of humanity as an entire whole. With regards to any issue, society tends to have two opposite choices, depending on perspectives, ways of calculating costs and benefits, the human capital of decision makers, information, and patrician divisions. Traditional saying that the friend of one's enemy is one's enemy, and the enemy of one's friend is one's enemy is applicable in party-based political games. Chinese Communist Party simply told the people: "whatever the enemy is against we should support, whatever the enemy supports, we should be against."

Complexity theory tells us that even disturbances owing to some trivial emotions can create different changes in society. Initial conditions may play a crucial role in choosing a specified evolutionary path among multiple possible ones. For instance, Japan and China were faced with similar threats from the Western powers in the early 1800s. The two rice-based societies evolved in different paths in their modernization. Utilizing the opportunity that the West had little interest in occupying the poor country, Japan soon modernized its army and economy. But, China failed to adopt an open policy and produced decreasing returns in front of Western (and Japanese) powers. Japan has experienced a linear process of Westernization, while China has had a highly nonlinear history of Westernization (socialism is also a part of Western creation). Depending on social and power structures, leaders tend to get biased information between positive and negative information. For instance, in China, the emperor, especially those after the emperor's family in power for a few generations, could hardly know objectively and correctly about what occurred in the society. This is also reflected in contemporary China which is dominated by a bureaucratic system run by smart, well-trained, and obedient members from the bottom of the society to the top. In such a system, some instructions of the top may be dampened or amplified by various levels of the bureaucratic system.

Nonlinear social systems have self-organizing capacities. It is difficult to predict its behavior because either external or internal, even very small, changes can lead to unpredictable consequences. For instance, Western intellectuals and politicians collectively failed to recognize the social and economic evolution of China from the 1990s to the early 2000s. After China has "suddenly" appeared strong in many fields, Western powers have been trying to adopt new strategies to deal with the new rising power. But timing is essential. Missed opportunity will not come back because China would never be what it had been even 40 years ago. Knowledge, memory, and environment are not irreversible. Collective stupidity or inefficiency is structurally similar, but does not necessarily lead to similar results.

Because it needs a lot of energy, a political system tends to be concerned with limited issues with some narrowly minded opinions (e.g., Cairney and Geyer 2017). The system accumulates political scum slowly and steadily if it does not have some self-correcting mechanisms until someday some sudden—either positive or negative—catastrophes take place. A rule controls some aspects but neglects others. This allows smart people to play with the rules to benefit themselves but harm society. Strange attractors imply that self-interested people may lead society for a long period without harmony and order. A society can be trapped in chaos and poverty. China's success in economic reforms was, fortunately, owing to the leadership of Deng Xiao Ping. A single man played a crucial role in saving a hundred million people from poverty—the greatest escape from poverty and ignorance in human history.

The dynamics occur within environments. Knowing yourself and knowing your environment is the basic requirement to make successful political decisions. The system self-organizes and the environment changes. Politicians may be slow to adapt to environmental changes. Many contemporary politicians are mocking stocks of the world not because they do not know themselves, but because they are ignorant about what is going on in many parts of the world. As changes can happen abruptly and decision makers may live in their comfort dreaming world, a great state may lose a weak state in competition.

The existence of multiple paths and the evolution of path dependence is important for understanding the difficulties of building a good society that requires a harmony of multiple forces. Path dependence implies that when a society has chosen a special ideology when it is faced with multiple ones (such as capitalism and socialism), it would be costly for it to switch to some more desirable ones. When an ideology or a faith is selected, society will make huge investments in constructing parts suitable for the ideology and destroying previous parts not suitable for the ideology. After some years, it would be costly to switch to other paths. The socialism under Chairman Mao is an example of the selection of the dominant thought. Chairman Mao destroyed the rich class (for his ideology of socialism) and destroyed China's higher education (owing to his Chinese peasant mindset in managing the country). When he was dying, China was at the very bottom of the world in terms of knowledge and per capita income. This made it quite easier for China to switch to another path of economic development. Some economies, such as India, which appeared far more promising than lowly educated China, for economic growth in terms of institutions, knowledge, and initial conditions in the 1980s, could not achieve economic miracles like China.

4.9 Strategy and Complexity Theory

A modern global company that employs many thousands of workers and produces hundreds of products is a highly complex system. Interdependence within an organization and interactions among organizations have caused the attention of researchers from different academic fields. Complexity theory enables scientists to look at organizational complexity from new perspectives. In organization theory, Anderson (1999) argues that complex adaptive system models enable researchers to simplify the complex by four key elements: agents with schemata, self-organizations sustained by importing energy, coevolution to the edge of chaos, and system evolution based on recombination. The organization is an open system because it exchanges resources with the environment. The organization is characterized by vertical hierarchy, horizontal complexity, and environment over time and space. Modern digitals enable an organization to simultaneously deal with problems of different parts of the Earth. A global organization is faced with political, cultural, and racial complexities. It is designed or evolves with adaption to the environment to survive in the local environment subject to global observation and rule. An organization can no longer compete in a global environment if it fails to adapt to environmental changes because it is faced with objective criteria in its status, quality, and price. The organizational structure must be designed to cope with its inner complexity and environmental complexity. Organizational inner complexity involves, for instance, human capital management. Labor supply is much dependent on the social environment which is beyond the organization's control.

An organization can hardly recruit highly qualified workers if the society is undeveloped in which few people receive formal education and experience an orderly social environment. Complex systems change inputs into outputs via multiple processes. A small error in one stage may lead to disasters in the output. Highly disciplined workers are required in each stage of the process. A usually simple job, such as doorkeeper, may require one to be educated because he must rapidly solve some unexpected but urgent problems. On the other hand, managers in organizations should reduce a complex problem to a simple description by omitting minor and unnecessary aspects. Many problems that seem intractable in designs are solved by practitioners. A modern manufacturing company from an undeveloped economy can hardly succeed in the global competition in a high-value-added industry. Underdeveloped economies do not provide a social, educational, and political environment for locals to survive in well-connected global markets. It is those companies from developed economies that grasp business opportunities.

Sun Zi's *Art of War* is the most well-known ancient book on the topic of strategy. It has charmed both military and business strategists in many parts of the world over centuries. From perspectives of complexity theory, the book is constructed rationally with *yin–yang* complexity. It expresses situation-dependent decisions and unexpected consequences even with well-prepared planning. During the war, there are no single well-defined cause-effect relations for the general to follow and very small changes in situations can lead to unexpected victories or defeats. Sun Zi asks for the absolute power of the general in his organization in war because it is through time-varying situations and the general's power in decision-making that an army can deal with unpredictable and quickly changing circumstances. Even the army has no fixed pattern or rules except that soldiers get disciplined and obey the order. Modern organizations are characterized by instability, adaptation, evolution, and structural changes. A few organizations can survive by keeping no structural changes with time in modern industrialized economies. As Schumpeter argues, endogenous change is a part of industrial evolution. Complexity theory implies that efficiency survival is sensitive to initial conditions, strategic actions, ability to enter the desirable attractor, and adapt or even lead the attractor of success. Chaos theory is extensively applied to the complexity of strategy. Levy (2007), for instance, explains how crucial managerial understanding and perception are important for making the right and innovative decisions for new challenges. Levy simulates the effect of managerial understanding on complexity in business. Chaos occurs when actors interact interdependently.

Alshammari et al. (2016) apply chaos theory to the complexity of organizations. Their research is focused on the resources need-motivated strategic interchanges, such as strategic alliances, mergers, and acquisitions. They emphasize initial conditions for the growth of companies. Initial conditions have an unpredictable impact and consequences on alliance formation and strategic alliance decisions. In the light of chaos theory, they argue that the engaged parties may be unaware of or neglect the small changes that take place anywhere and anytime owing to the organizational cultures and other external environments until some big changes take place.

Firms ally putting alliances' resources to create value. The alliances start with initial conditions for formatting partnerships to utilize mutual resources and knowledge. Each firm may have a different status in competition, formal and informal relationships with society and other firms, groups of customs, locational advantages, and the like. When the alliance is made, each player in decision positions would predict short-term as well as long-term costs and benefits, and consequences. Initial conditions are often uncertain, unstable, and even deceptive. This implies that the initial decisions are chaotic in the sense that the outcomes are not precise or do not follow the predictions of the partners involved. It is observed, for instance, that when international acquisitions are concerned, cultural, political, institutional, and language differences are important factors to consider (Owen and Yawson 2015). These differences affect knowledge transfers and mutual understanding. Some tangible or intangible resources become useless. Speed and costs of acquisition will be affected.

Through merger and acquisition, the key skills or techniques can be secured. There are many factors for the results of merger and acquisition. Tax reforms, number of buyers of the target firm, bidder's approaches, mode of payment cash or stock, and the like. There is no single key factor identified by the research on the topic over the years. Firms always struggle to capture every possible potential for advantage. "Competing-on-the-edge strategy" is the way to survival and even to a superior position. Since, firms try to get an advantage by collecting information, employing the best workers, accessing to best technologies and making innovations, accepting deceptive as well as cooperative strategies, securing resources, occupying the market, using political as well as other social and business networks, the business world is chaotic and unpredictable. The strongest few, in some industries, occupy the chaotic attractor which other players can only watch or try hard to share something left by the dominant attractor. Alshammari et al. (2016) pointed out: "The four possible courses of movement in for firms concerning to the chaos theory are: first. A fixed-point attractor would attract the system, at which the system will stabilize and move like a plumb until the complete stop before starting a new phase. Second. Sporadic attractor through which the system is prone to cyclic behavior. Third. The number of patterns of behavior increases as the system behavior starts repeating itself. And lastly, the strange attractor where the system can never repeat a certain behavior after having been looped into the strange attractor space." If it has no absolute advantage in competition and/or innovation, a firm tends to be allied with other firms and a new business structure may emerge. Changes take time. The alliance may also bring about responsiveness and an inability to deal with new situations and unpredictable happenings. Before the alliance can perform better, markets may already be occupied by new entrants.

4.10 Geography and Nonlinear Cultural Evolution

Geography is a difficult dimension for sciences to deal with. Mathematically, spatial dynamics lead to partial differential equations which are solvable only in limited situations. Cultural sophistication and falsification are long-term processes limited to space. Change over space is essential for cultural evolution. In astronomy, chaos theory gives new insights into the analysis of globular clusters, the huge star groups akin to the Milky Way which make up most of the universe. A globular cluster is a many-body problem. The two-body system was solved by Newton. The Earth and the moon, for example, travel in a perfect ellipse around the system's joint center of gravity. The three-body system is often incalculable. Orbits can be tracked for a time, but chaos soon characterizes the calculation. Even if one knows the physical principles of the movement of stars, one cannot follow the consequences of interactions between stars. This is also true of personal complexity, regional complexity, racial complexity, as well as humanity over space. As humans are connected, the decline, rise, and perish of any powerful family, region, nation, or race is not a low probability event.

Immanuel Kant (1724–1804) paid great attention to geography. Kant was born at Konigsberg in East Prussia. He is often considered the greatest philosopher among Germans. He made great contributions to the fields of geography, astronomy, and geology. Kant made an important contribution to geography, even though he did not publish. He gave lectures on geography for almost forty years. His lecture notes were circulated among students. His writings, published as *Physische Geographie* in 1802, are his lecture notes. He holds that geography and anthropology provide knowledge of the world (Elden 2009; Elden and Mendieta 2011). His contributions to geography are not made via investigation, and observations of world geography, but by laying the philosophical foundation of the discipline of geography. During the long period from 1756 to 1796, he gave a regular course on geography. According to Kant, the research of geography is an essential approach to empirical knowledge. Geography plays a pivotal role in the process of human civilization. Kant considers natural geography and geography by human actions. The study of history is focused on time, examining phenomena that follow one another in order of time, while the study of geography is concerned with space, dealing with phenomena that spread beside each other in space. Any Individual's experience occurs over a specified space and time. Kant believes that space is something objective or real, it is something subjective or mental. It is governed by an unchanging law. Space provides a kind of framework for coordinating things and events all of which can be perceived with the outer senses. The entire gamut of empirical knowledge, which can be classified according to time and space, is composed of the study of history and geography. For Kant, about whether geography or history exists first, they simultaneously coexist. Each event occurs in specified history and geography, which implies that each culture is unique. There is no generalized knowledge about man and society. Germany is German, China is Chinese, and America is American. They all have exceptional features. This differentiates the study of culture from natural sciences. Cultural and historical sciences study those objects that are mainly mental constructs. Kant also developed different branches of geography, such as mathematical geography, moral geography, political geography, commercial geography, and theological geography.

In modern times, geography has become interdisciplinary There is no clear-cut division between the sciences as argued by Kant. Hegel (1956) argues:

> The Greeks and Romans had reached maturity within before they directed their energies outwards. The Germans, on the contrary, began with self-diffusion-deluging the world, and overpowering in their course the inwardly rotten, hollow political fabric of the civilized nations. Only then did their development begins, kindled by a foreign religion policy, and legislation" (p. 341)

Shapiro (2015: 318) explains:

> The very being of the German people is their transformation through encounters with the other, so they are uniquely suited to confirm Hegel's concept of the true identity as the identity of identity and non-identity. They seize Rome and appropriate Christianity almost thoughtlessly, but such is the cunning of history - they are transformed in the end by what they have captured. They are predatory subjects who will be transformed by their object. ... Their wandering, migration, and nomadism become subordinated to the process of state formation in which religion is essential (p. 318).

The 2008 Nobel Memorial Prize in Economic Science is awarded to Paul Krugman for his contributions to new trade theory and new economic geography. The Prize Committee claims that his works explain the patterns of international trade and the geographical distribution of economic activity, emphasizing economies of scale and consumer preferences for diverse goods and services. Spatial distribution is related to agglomeration and dispersion forces. New economic geography reveals mechanisms and phenomena of spatial location with different mechanisms and factors (Zhang 2008; Commendatore et al. 2015). There is a huge amount of literature with applications of different economic mechanisms and mathematical techniques.

Returns to scale, scope, and networking economies are the key concepts to describe contemporary, especially developed economies. Digital technologies and low-cost and rapid modern transportation dramatically increase the impact of these economies on economic geography. Megacities are expanding without any sign of the end. Global innovation is concentrated in a few centers. Enlarging spatial differences in productivity, incomes, creativity, innovation, infrastructures, and growth are largely associated with these economies (Bettencourt et al. 2007; Proost and Thisse 2019). Increasing returns to scale implies that if a region doubles its inputs of production, its GDP is more than doubled. A few smart entrepreneurs, in the IT industry, can bring about great changes in a city. Increasing returns to scale explain much about the sudden rise of China and spatial concentration and dispersion of industrial production, innovation, concentrations of services, and banking systems. Increasing returns is not only owing to innovation and production but also owing to consumption (Bond-Smith 2021). Rising powers in the political and business worlds of India and China are not so much owing to their political ideologies or creativity as owing to the size of the population. It is easy to see that decreasing returns to scale, for instance, owing to fast population growth in some poor regions will keep them in poverty tarps for quite a while.

A high proportion of knowledge is almost freely available to a high proportion of the global population. Population growth has an uncertain interdependent relation with knowledge stock. Knowledge is not free because it requires effort. Accumulating knowledge and applying knowledge are often separate sources and processes, even though they often occur simultaneously. A firm has most of its inputs commonly used by all other firms in the same industry but becomes the monopoly in the long term because of its unique advantage in some input factors and increasing returns. The number of unskilled and uncreative people has little to do with enhancing productivity but plays a key role in the demand side. In economic competition, mankind comes to a stage in which the masses play a key role not so much in production as in consumption. The rich's wealth requires the masses' dreams and consumption, and the power requires the crowds' obedience and trust. When we come to studying national levels of competition, nonrival factors such as cultural reputations and institutions play an important factor in securing power from increasing returns to scale.

Life is a process of construction through consciousness and collectively accumulated objective knowledge and emotional expressions under largely "given" environments. Man can individually achieve almost nothing if not born into a certain proper environment for personal growth. A great American businessman might be just a smart peddler on an Indian street if born in India or a diligent land-digger farmer in Culture Revolution China. Many richest men in China—who were so lowly educated and had no rich experience in anything but poverty—might have difficulties finding any proper job in mainland China if they were born 30 years later. Similarly, a race with great achievements might achieve nothing interesting when timing (fortune) is not proper. In the last century, for instance, Jewish people in the West and Hakka people in Asia made great achievements in globally liberating and chaotic environments. Poor and low-status immigrants who had almost no knowledge about higher elements of their own cultures like the United States or devasted countries after the war like Japan and Germany had provided suitable conditions for fast economic growth under the international conditions and technological levels. Timing and space play essential roles in personal and cultural growth. Complexity theory provides new insights into the importance of timing and space in the self-organization of humans as well as human societies.

A civilization is characterized by its current status and stocks of beliefs, arts, music, customs, technologies, and knowledge accumulated over centuries. Civilization is a super complexity beyond any knowledge to fully describe. Its process is nonlinear and full of chaos, bifurcations, structural changes, and catastrophes. No civilization is ever glory long. Perhaps, no civilization is ever decent in human history. The deep dark side of civilization is not told by itself, and no other can display it truly. It is created simultaneously via combinations of conscious and unconscious forces, rational and irrational thought, predictable and unpredictable factors, intuitive insights and logical reasoning, and the like. Civilization does not display any certain patterns or paths. It might die young even before it matures once. It might also achieve different golden periods and experience slavery periods.

East Asian people, mainly Japanese, Koreans, and Southern Chinese (and emigrants from the South to various parts of the world) are often called "cunning" in the West. This cultural stenotype might not be ill-grounded. If one explores these people with a Newtonian vision of the world, one will get confused because this region interprets life in the *Yi Jing* or yin–yang worldview. In Chinese society, the complexity of human relations increases exponentially when more people are tied together. East Asia has the highest average IQ on Earth. This is owing to the complexity of human networks in these societies. People were fixed on land and the family lived in the same village over many generations. A man in charge of the family was responsible to deal with many hundred relatives. To remember each of these people and their relations with other people and properly deal with them, following customs within the budget (plus masting the Chinese language), requires a village leader's energies and intelligence as much as professors in a modern mediocre university.

Human action takes place in geography. Humans have shaped geography and geography has shaped humans. Love unites and hates separates. Hatred often occurs between neighboring states. As modern technologies have transformed the role and power of geography in human relations, modern trust between states is no more bound by geography. Geography has a lasting impact on a sensitive soul—this is reflected in songs of memorizing home landscapes. From climates, soils, and animals with which culture is mingled, we can find some roots of the cultural formation. Even after they have been culturally, economically, and politically connected, differences in cultures and human capital structures between Italy and Sweden are characterized by their geographical differences. George Santayana said: "For me, it seems a dreadful indignity to have a soul controlled by geography." A sentimental heart reflects geography. Unconscious reflections and emotional reckoning back to the geography that one had been brought up with. Neil Gaiman (1984), an English author, says: "The biggest difference between England and America is that England has history, while America has geography." Values, like mathematics and physical laws, are universal. But each universal value is manifested variedly across the Earth. Complexity theory perceives human affairs consequences of complicated interactive networks. Trust and justice are closely related to geography, religion, economy, culture, and technology.

Between the inside and outside of a system, there is a boundary. To comprehend the contemporary world, one needs to consider "boundaries" as changeable and endogenous parts of global evolution. Cilliers (2001) describes:

> Boundaries are simultaneously a function of the activity of the system itself and a product of the strategy of description involved. ... The boundary of the system is therefore neither purely a function of our description nor is it purely thing…. [N]on contingent subsystems could be part of different systems simultaneously. This would mean that different systems interpenetrate each other, that they share internal organs (pp. 141–2).

Schopenhauer points out: "If a man does not love solitude, he will not love freedom; for it is only when he is alone that he is really free." Sigmund Freud (1856–1939) wraps up his analysis of the psychology of man: "The liberty of the individual is not gift of civilization. It was greatest before there was any civilization." Each civilization has its scope and degree of individual freedom. Freedom of society has historically been formed by geography. Rice paddy, nomadic, desert-based, island, and agrarian people have created, applied, and sustained different views on freedom, human relations, and religions (or beliefs) and have varied customs in processes of coexistence with nature and fighting against their neighbors. Geographical influences on races are printed in behavior and social organizations. For instance, nomadic culture is bolstered by respecting rule, kindness, and democracy within the group and cruelty to outsiders. This is due to economic survival and the dynamics of group formation. Before the West opened China, China knew two basic types of societies. One is rice paddy based. The typical image of such a society is immobile, connected by cultural customs, obedience, and cruelty within the group for conformity. The other is nomadic. The image is mobile and connected with democratic-like sharing benefits and slaughtering the outsider without rule, not to mention the law. The sustainable development of the rice society is by—consciously or unconsciously—creating and maintaining a set of cultural values by which it suppresses free expression of emotion, makes decisions based on detailed calculation of costs and benefits with the long-term relationship as a calculating basis, has no freedom of sexuality, is extremely cautious in accepting the outsider, has no tradition of obeying simple and well-developed law or rule, has an advanced culture as the basis of class division, and has a dominant rational or irrational thought system as the dominant bolster of organization, accepts multiple religions and thought systems for coexistence of well-cultivated minority, has a fixed home with land as the lasting foundation. The nomadic society, if its population is large, is united by targeting a common wealthy enemy. Wealth, women, wine, and a comfortable life are the common goals for men. Men would have unlimited pleasures in preparing for, planning, fighting for, and winning battles. Nomadic people of multiple groups have some primary beliefs and boundless plateau. They don't harshly suppress the expression of emotion, decide on the calculation of costs and benefits with relatively short-term relations, have no strict control on gender relations, are "quick" to recruit outsiders to form a stronger and larger team, have simple and clear rules, has a merit-based group order, and has mobility and flexibility. In such societies, the mighty is the source of wealth, and the mighty is the source of sex. Any spirit can go as high as possible in nomadic cultures. In *On Liberty*, Mill compares the fall of China and the rise of Europe as follows:

> We have a warning example in China - a nation of many talents, and in some respects, even wisdom, owing to the rare good fortune of having been provided at an early period with a particularly good set of customs, the work, in some measure, of men to whom even the most enlightened European must accord, under certain limitations, the title of sages and philosophers. ... On the contrary, they have become stationary... They have succeeded beyond all hope in what English philanthropists are so industriously working at - in making a people all alike, all governing their thoughts and conduct by the same maxims and rules; and these are the fruits.
>
> What has made the European family of nations an improving, instead of a stationary portion of mankind? Not any superior excellence in them, which, when it exists, exists as the effect not as the cause; but their remarkable diversity of character and culture.... Europe is, in my judgment, wholly indebted to this plurality of paths for its progressive and many-sided development.

Different from China, Europe has never built a unified Europe because no region could conquer all the other areas and each region has distinctive cultural and political features. In *War and Peace*, Leo Tolstoy (1828–1910) illustrates:

> Germans are self-confident based on an abstract notion- science, that is, the supposed knowledge of absolute truth. A Frenchman is self-assured because he regards himself personally, both in mind and body, as irresistibly attractive to men and women. An Englishman is self-assured, as being a citizen of the best-organized state in the world, and therefore as an Englishman always knows what he should do and knows that all he does as an Englishman is undoubtedly correct. An Italian is self-assured because he is excitable and easily forgets himself and other people. A Russian is self-assured just because he knows nothing and does not want to know anything, since he does not believe that anything can be known.

The traditional European variation was in contrast to China's situation. With traditional weapons and European geography, it was challenging for any hero to conquer different parts. State competition and other factors led to advances in technology (and weapons). East Asia had been far behind Europe when the West forced China and Japan to open their doors. Japan's poverty and geography played an important role in its modern successes.

Adam Davidson points out: "Poverty is not the simple result of bad geography, bad culture, bad history. It's the result of us: of the ways that people choose to organize their societies." One is aware of the complexity of the economic landscape when watching from an airplane over metropolitan regions and traveling on trains. Economic geography refers not only to what one sees physically but also to cultural and monetary values. Tangibles and intangibles intertwine over space to sustain a world beyond anyone to comprehend. Profit-hunting companies, utility-maximizing households, and welfare-maximizing governments jointly create collectively inscrutable spatial patterns over time. To travel is to appreciate the complexity of geography. Home, family, national territory, cultural identity; all these emotional words are referred to geography. Geography has a decisive force on cultural formation before modern times. Technological changes have been dismantling traditional barriers of time and space on global cultural formation and "destroying" many cultures constructed over a thousand years under special geographical conditions. Mankind is interacting with changed geographical conditions. Interdependence between human survival and geography brings about a new macro level of complexity that mankind has neither knowledge nor experience. The world displays a complicated network of relatedness between products, or the product space as termed by Hidalgob et al. (2007). Hidalgob et al. map that most upscale products are located in a densely connected core, while low-scale products are distributed across a less connected periphery. Countries tend to move and locate goods near to those they are currently specified in, taking advantage of imitation, learning, and access to effective resources. This implies that once a country enters the poverty trap, it requires some "fortunes" and special efforts to escape from poverty because it would be very hard for poor economies to develop very competitive exports with higher value-added components. Agricultural products and simple (or low-skill, labor intensive) components of high-value-added products cannot enable the poor to converge with the rich. Forces of agglomeration have led to a main feature of the global product that is modular with some goods highly connected and others disconnected. This is similar to the academic field of intellectuals, especially those engaged in mathematics and natural sciences. Money (reputation) loves and flows to those who love and care about it. Over time, it does seem that the global product space is becoming sparser. It tends to display a core-periphery structure. The core is concentrated with high value-added economic activities, while the periphery is characterized by the rest of the products.

One complexity of a spatial economy is the location of production and consumption of products across the Earth. The modern world produces a few or millions of kinds of products and services, depending on to what scale one makes classification. The complexity of products, from airplanes, computers, shoes, and apples and from higher-value-added to lower value-added products, is varied among products. Scale, scope, agglomeration, diffusions, accessibilities to knowledge, technologies, cultures, institutions, costs of technologies, labor, lands, and taxes, are all factors interacting with each other in producing, for instance, computers and cars. Civilizations were once clustered owing to global geographical isolation and local leveraging of ideas, technologies, and skills through conflicts and cooperation. Globalization divides the world rapidly into the poor and the rich. Except for a few resource-rich regions, rich economies tend to produce higher-value-added products (processes) with sophisticated technologies, available resources, a highly skilled labor force, and a stable social and institutional environment. While poor economies tend to supply lower-value-added products (processes) with relatively simple technologies. In a globally connected world, the rich are similarly rich via extensive imitation, innovation, learning, and adaption, while the poor are poor in their survival logic.

Leibniz correctly predicts that someday China would be invaded by Europe because the Manchu government neglected advancing technology and weapons. In 1697, Leibniz observed the reasons for the Manchu dynasty to be conquered by the West:

> They [the Chinese] also yield to us in military science, not so much out of ignorance as by deliberation. For they despise everything which creates or nourishes ferocity in men, and almost in emulation of the highest teachings of Christ…, they are averse to war. They would be wise indeed if they were alone in the world. But as things are, it comes back to this, that even the good must cultivate the arts of war, so that the evil may not gain power over everything. In these matters, then, we are superior.

The West's ideal was for an open world, while the (non-Han) Manchu emperors' vision was for an isolated one. The Manchus had absolute power and advanced weapons within the dynasty. They did not need to put more resources to develop weapons because that might endanger the Manchu power. The spread of more advanced weapons among neighbors or the Chinese would jeopardize the Manchu domination over China. One hundred years later, the West equipped with big ships and advanced weapons gave the Manchu court painful lessons. China is a rice paddy society under a centralized power over a vast territory surrounded by nomads whom Han Chinese could not conquer or control before modern weapons were available. Before modern weapons had been introduced to China, the Chinese empire's main problem was to construct and maintain the Great Wall, rather than expand its territory beyond the Great Wall. A divided China would invite nomadic invasions. If Han Chinese were not led by a single centralized power, China would hardly be peaceful, especially after nomadic peoples had mastered more advanced strategies through many generations of learning by fighting. Great power is bolstered by forces and opportunities. Mongols dominated China owing to their strong body, team spirit, flexibility in organization, speedy horses, and killing skills. Mongolian courage was gone long ago in the front of the Chinese (Manchus) after Manchus held effective cannons and mindsets of understanding both the rice-based and nomadic civilizations.

China has almost all kinds of climates, which explains why Chinese food has such a vast variety. The hot climate plus the varieties of the geography of Southern regions gives China a global reputation that Chinese people eat whatever is available. In most Chinese Western and Northern regions, there was not a great variety of traditional food because many kinds of vegetables, fish, animals, and birds were not available. In Europe, France, Italy, and Spain are great places for hunting food, while Scandinavia, Britain, and Germany have some but not many varieties of traditional gourmets. These diversities are determined by geography. Montesquieu (1689–1755) observes Chinese characters a long time ago:

> Power in Asia ought then to be always despotic: for if their slavery was not severe, they would soon make a division, inconsistent with the nature of the country.

China's geography is important for the formation of Chinese obedience, while Europe's geography explains much about its diversity and leadership in modern civilizations.

An important index of national human capital formation is IQ. There are different studies on IQs among nations and regions. Ranking results are similar. Geography plays an important role in determining IQ differences among nations. As listed in Table 4.1, China and its neighbors (which are connected even since ancient times) have similar levels of IQs. In Europe, France, Germany, and England have similar IQ levels. High-IQ countries in East Asia are also well-known for being lowly ranked in happiness. The table lists the average IQs for some selected countries. The data for 110 countries was made by a project called WorldData.info. The IQs in the table were averaged from the results of 9 international studies. IQ is a measurement of ability for combinations, comprehension, and learning. The average IQ of an economy seems increasingly important in an information-based and knowledge-based economy.

Human geography, trust, justice, economy, and other factors are simultaneously active. It is the brain that separates and isolates them. Mark Kurlansky (1948), an American journalist and writer, suggests:

> As with wine, geography affects the flavor. Oysters are usually named for a locale. … Food is the best way to teach history and geography and most everything else.

Each human being is a (unique organic) complexity. Each typical local dish is a complexity that can hardly be exactly repeated somewhere else. One who stays in a small town for a lifetime can become a great thinker who finds some concrete law of the universe from an unlimited variety. A man who loves to obverse human variety tends to love to travel; a man who loves food tends to hunt for food. Benjamin Gibbard (1976), an American singer and songwriter, reflects:

> When you connect as many memories to your geography as I have, and then you see that geography changes around you, you're forced to reckon with time.

What Gregory R. G. Gianforte (1961), an American businessman and the 25th governor of Montana since 2021 concludes, perhaps from a technological perspective, "Once the internet removes geography as a constraint, the smartest people go to the most beautiful places." Culture is an organic complexity and man is filled with unconscious mystery ideas and emotions. Brooke Baldwin (1979), an American journalist and TV host at CNN (2008–2021) voices:

> What does 'home' really mean? Is it merely geography, where you were born? Could it include straddling two continents and cultures? Or perhaps it's a place with a spiritual magnetism – a feeling toward a culture or people – that's tough to put into words?

Max Weber (1864–1920) expresses: "'Culture' is a finite segment of the meaningless infinity of the world process, a segment on which human beings confer meaning and significance." It is easy for contemporary readers to comprehend what this great German sociologist concludes. An interesting human life does seem, to be characterized by searching for the meaning of life. If human life has no meaning at all for human society (including himself), life itself is nothing but a combination of something that man still does not know with certainty.

4.11 On Complexity of Wars

To effectively kill for some faith or territory without getting killed is a royal road, at least for certain personalities, to personal glorification, economic security, and high social status. American economist Thomas Sowell reveals: "There are only two ways of telling the complete truth—anonymously and posthumously." No other human affair in human history is so complicated as wars. Humans devote all available resources—such as human lives, animal lives, emotions, national resources, wealth, skills, spirits, technologies, knowledge, and strategies—to win or prevent wars over history. China's Great Wall cost incalculable lives and national resources over a thousand years to prevent nomadic invasions. Traditional human societies evolved under conditions that a high proportion of males die in wars. This has a significant implication for the stability of gender markets. Modern advanced economies are faced with two new phenomena: young men do not die in war and life expectancy becomes much longer. New socioeconomic structures are needed for domestic social stability and sexual satisfaction. Still, a high proportion of young people is daily trained and prepared to kill other humans. The costs of maintaining a modern army are high but the efforts are luckily wasted. Trained men get their hearts hardened. Some of them turn out enemies of their society when they are not treated properly after completing military service. Nations often need wars to solve their domestic problems. There are always interest groups who would profit greatly from wars. These kinds of people in modern times would find all possible reasons to start wars if they do not fight and don't have risks of losing a life. Some people get direct benefits from fighting such as social glorification and pleasure. Benefits from wars are still high but glorification as being human from wars is greatly diminished.

Social change is closely related to technological change. Civilization advances. In every new epoch of technology, man kills man in more effective ways. Mongols were once strong on the Earth because they had strong bodies, fighting spirits, and skills, as well as advanced weapons. British conquered most of the Earth with their simple—in modern standards—weapons and ships. America could not become a superpower without advanced ships and nuclear powers after WWII. Blaise Pascal tried to find justice in wars: "Can anything be stupider than that a man has the right to kill me because he lives on the other side of a river and his ruler has a quarrel with mine, though I have not quarreled with him?" But, this is what mankind has done from some unknown time. Human interactions are complex. The right is defined and manipulated by the mighty or rich over history. Intellectuals are collectively seduced or forced to serve the rich and the powerful. Those who could honestly and intellectually tell the truth about man and society and happily survive long are not normal but rarely lucky among human races. WWII was the latest—perhaps, not final—war between equal great powers. Aldous Huxley reveals a cruel fact about persistent humanity:

> The most shocking fact about war is that its victims and its instruments are individual human beings and that these individual beings are condemned by the monstrous conventions of politics to murder or to be murdered in quarrels not their own.

Ernest Hemingway knew about the consequences of war in modern times:

> There is no hunting like the hunting of man, and those who have hunted armed men long enough and liked t, never care for anything else thereafter.

Man is a complex social animal with capacities of reasoning and infinite imagination for future events. Men are culturally mainly made of phantasies. Conflicts are owing to different interests and people innovate different reasons, emotions, and methods to profit themselves. Wars occur between societies as an effective and beneficial way to solve conflicts. In the territory of contemporary China, races, and states appeared and vanished as a consequence of wars over many thousands of years. No one knew about how many languages and races existed in the territory. George Orwell argues: "War against a foreign country only happens when the moneyed classes think they are going to profit from it." Khalil Gibran asks: "Where is the justice of political power if it executes the murderer and jails the plunderer, and then itself marches upon neighboring lands, killing thousands and pillaging the very hills?" John Stuart Mill reasons the war in a typical British colonist mindset: "War is an ugly thing, but not the ugliest of things. The decayed and degraded state of moral and patriotic feeling which thinks nothing is worth war is much worse." Human history is not a history of justice but a history of mighty progresses and wars. War connects people and contributes to human progress. Considering the number of people right now sustained on the Earth, we might positively think that men are living in a prosperous and relatively safe world as a consequence of frequent wars across geographical boundaries.

One group might imagine its god and use the name of the god to conquer and kill others. Napoleon Hill holds: "War grows out of the desire of the individual to gain an advantage at the expense of his fellow man." Ernest Hemingway characterizes war gaming: "Once we have a war there is only one thing to do. It must be won. For defeat brings worse things than any that can ever happen in war." War is a passionate disease. Monkey societies provide a much clearer illustration of collective behavior. Unexperienced youths passionately fight for the causes of the war they have little understanding of the costs of their lives. A great American fighter G. S. Patton glorifies war:

> Battle is the most magnificent competition in which a human being can indulge. It brings out all that is best; it removes all that is base. All men are afraid in battle. The coward is the one who lets his fear overcome his sense of duty. Duty is the essence of manhood.

It is significant for man to be dutiful even for the survival of a social group. A question not yet unanswered is "duty for what." Should the target of the male duty be for the woman, or the family, or the group, or the club, or the nation, or humanity? The answer varies among individuals and societies over time. A good man should be dutiful for everything, as a beautiful and valuable woman might ask for. But, it does seem true that multiple duties kill manhood. Deception is an economical way to escape from duty burdens.

Direct wars between the powerful have not occurred for over 80 years perhaps because of destructive forces of nuclear powers. Otto von Bismarck believes: "The great questions of the day will not be settled using speeches and majority decisions but by iron and blood." But wars continue in the way that Winston Churchill predicts: "When the war of the giants is over the wars of the pygmies will begin." Similarly, when the nation enjoys long-term peace and prosperity, conflicts between women and men will and divorce rates increase. Politicians play wars for their benefit in the name of national benefits. The rich avoid direct fighting on battlefields with their heart focused on profits. In modern times, it is youth, often from families of low social and economic status, who risk lives without being aware of the meaning of war and its consequences for themselves. American John K. Galbraith admits his own country's behavior: "War remains the decisive human failure." War was once a royal road to nobility for man. Paul Valery describes the obvious: "War: a massacre of people who don't know each other for the profit of people who know each other but don't massacre each other." There is no solid reason to kill other humans for nothing truly meaningful. It sounds always admirable and encouraging if one claims to die for one's territory. Monkey society naturally supplies a high proportion of soldiers. This is the natural design of the evolution of selfish genes. Rational thinkers give similar solutions to end the war as Thomas A. Edison suggests:

> There will one day spring from the brain of science a machine or force so fearful in its potentialities. So absolutely terrifying, that even man, the fighter, who will dare torture and death in order to inflict torture and death, will be appalled, and so abandon war forever.

Nietzsche points out: "The best weapon against an enemy is another enemy." An adult man easily finds another man as an enemy but hardly has a genuine friend. He can luckily find a dog as his best friend and loyal companion. Modern people collectively learn much from dogs: smile at and play with man to get served by man.

Table 4.1 Average IQs, GDP per capita, and happiness indexes

Country/region	IQ rank (IQ)	GDP per capita (US$) in 2017	Happiness rankings
Singapore	1 (108)	57,713	34
Hong Kong	2 (108)	46,194	36
Taiwan	3 (106)	n.a	26
South Korea	4 (106)	29,742	57
Japan	5 (105)	38,428	54
China	6 (104)	8,826	86
Switzerland	7 (102)	80,190	5
Germany	15 (100)	44,769	15
United Kingdom	16 (100)	59,720	19
France	26 (98)	38,477	23
United States	27 (98)	59,532	18

Sources The IQs: https://www.worlddata.info/about.php
The GDP in 2017: https://data.worldbank.org/indicator/NY.GDP.PCAP.CD
The 2018 happiness index: https://en.wikipedia.org/wiki/World_Happiness_Report

But a man is still human; man cannot persistently and freely smile at each other like a dog from the bottom of its heart (who knows?).

References

Alshammari M, Pavlovic M, Qaied BAA (2016) Chaos theory in strategy research. American Journal of Business and Management 5: 1–13

Anderson P (1999) Complexity theory and organization science. Organization Science 10: 216–232

Arrow KJ (1962) The Economic implications of learning by doing. Review of Economic Studies 29:155–173

Beeren, C von (2012) Social integration of macro-parasites in ant societies: Proximate and ultimate mechanisms. Südwestdeutscher Verlag für Hochschulschriften

Bell, D (1996) Afterword: 1996, in The Cultural Contradictions of Capitalism. New York: Basic Books

Bernays E (1928; 2004) Propaganda, with an introduction by Miller, C. Ig Publishing, Singapore

Bettencourt LM, Lobo J, Strumsky D (2007) Invention in the city: Increasing returns to patenting as a scaling function of metropolitan size. Research Policy 36: 107–120

Bond-Smith S (2021) The unintended consequences of increasing returns to scale in geographical economics. Journal of Economic Geography 21: 653–668

Bourdieu P (1998) Practical reason: On the theory of action. Stanford University Press, Stanford, CA

Buckley EP, Watson JL (1986) Kinship organization in late imperial China 1000–1940. University of California Press, Berkley, CA

Cairney P, Geyer R (2017) A critical discussion of complexity theory: How does complexity thinking improve our understanding of politics and policymaking? Complexity, Government & Networks 3: 1–11

Cairney P (2012) Complexity theory in political science and public Policy. Political Studies Review 10: 346–358
Cilliers P (2001) Boundaries, hierarchies and networks in complex systems. International Journal of Innovation Management 50: 135–147
Commendatore P, Kubin I, Sushko, I (2015) Typical bifurcation scenario in a three region identical new economic geography model. Mathematics and Computers in Simulation 106: 63–80
Cowan RS, Hersch MH (2017) A social history of American technology, 2nd. Oxford University Press, Cambridge
Elden S, Mendieta E (2011, eds.) Kant's geography. Suny Press, New York
Elden S (2009) Reassessing Kant's geography. Journal of Historical Geography 35: 3–25
El-Ghalayini Y (2017) Complexity theory: A new way to look at new public management. Network and Complex Systems 7: 6–10
Elias N (1992) Time: An essay. Wiley-Blackwell, Oxford
Fei CH, Liu TJ (1982) The growth and decline of Chinese family clans. Journal of Interdisciplinary History 12: 375–408
Ferguson-Gow H, Sumner S, Bourke A, Jones KE (2014) Colony size predicts division of labour in attine ants. Proc. R. Soc. B281:20141411. https://doi.org/10.1098/rspb.2014.1411
Fiese BH, Winter MA (2010) The dynamics of family chaos and its relation to children's socioeconomic well-being, in Evans GW, Wachs TD (eds.) Chaos and its influence on children's development, pp. 49–66. American Psychological Association
Foroohar R (2017) Makers and takers – How Wall Street destroyed main street. Crown Business, New York
Friedmann J (2007) Reflections on place and place-making in the cities of China. International Journal of Urban and Regional Research 31: 257–279
Fuchs C (2003) Structuration theory and self-organization. Systematic Practice and Action Research 16: 133–167
Gell-Mann M (1994) The quark and the jaguar: Adventures in the simple and the complexity. W.H. Freeman, New York
Greif A, Tabellini G (2010) Cultural and institutional bifurcation: China and Europe compared. The American Economic Review 100: 135–140
Gordon D (1999) Ants at work: How an insect society is organized. Free Press, New York
Gordon D (2010) Ant encounters: Interaction networks and colony behavior. Princeton University Press, Princeton
Green DG (2014) Of ants and men: The unexpected side effects of complexity in society. Copernicus, Gottingen
Greenspan A (1996) The challenge of central banking in a democratic society, the speech delivered at AEI's Francis Boyer Award on December 5, 1996
Haken H, Kelso JAS, Bunz H (1985) A theoretical model of phase transitions in human hand movements. The Biological Cybernetics 51: 347–356
Harrington JL (2008) Technology and society. Jones& Bartlett Learning. Mass., Sudbury
Harvey D (2001) Spaces of capital. Routledge, London
Hegel GWF (1956) The philosophy of history, translated by Sibree JD. New York
Hidalgob CA, Klinger B, Hausmann R, Barabasi AL (2007) The product space conditions the development of nations. Science 317: 482–487
Hughes J, Cain LP (1998) American economic history. Addison-Wesley, New York
Jullien F (1995) The propensity of things toward a history of efficacy in China, translated from the French origin by Lloyd J. Zone Books, New York
Jung CG (1950) Foreword to *The I Ching*, or *Book of Changes*, translated by Richard Wilhelm. Princeton University Press, Princeton
Jung CG (1997) Jung on synchronicity and the paranormal, key readings selected and introduced by Main, R. Routledge, London
Kelso JAS (1995) Dynamic patterns: The self-organization of brain and behavior. The MIT Press, MA, Cambridge

Leiserson G, McGrew W, Kopparam R (2019) The distribution of wealth in the United States for a net worth tax. The Washington Center for Equitable Growth, March 21, 2019. Retrieved from: https://equitablegrowth.org
McClosky H, Zaller J (1984) The American ethos: Public attitudes toward capitalism and democracy. Harvard University Press, Mass., Cambridge
McGee ZA, Jones BD (2019) Reconceptualizing the policy subsystem: Integration with complexity theory and social network analysis. Policy Studies Journal 17: S138–S158
Myers HA (1982) Western views of China and the Far East - Ancient to early modern times. Asian Research Service, Hong Kong
Nooten SS, Schultheiss P, Rowe RC, Facey SL, Cook JM (2019) Habitat complexity affects functional traits and diversity of ant assemblages in urban green spaces: Formicidae. Myrmecological News 29: 67–77
Ortiz-Ospina E, Roser M (2020) Marriages and divorces. Retrieved from: https://ourworldindata.org/marriages-and-divorces
Owen S, Yawson A (2015) R&D intensity, Cross-border strategic alliances, and valuation effects. Journal of International Financial Markets, Institutions & Money 35: 1–17
Parker G (2008) Crisis and catastrophe: The global crisis of the seventeenth century reconsidered. The American Historical Review 113:1053–1079
Proost S, Thisse JF (2019) What can be learned from spatial economics? Journal of Economic Literature 57: 575–643
Redding SG (1993) The spirit of Chinese capitalism. 2nd edition. Gruyter, Berlin
Sabadini SA (2019) Riding the Waves of Change – An introduction to the I Ching. Retrieved from https://www.meer.com/en/56314-riding-the-waves-of-change
Saperstein AM (1999) Dynamical modeling of the onset of war. World Scientific Publishing, Singapore
Shapiro G (2015) World, earth, globe: Geophilosophy in Hegel, Nietzsche, and Rosenzweig. Iyyun: The Jerusalem Philosophical Quarterly 64:313–334
Simler K, Hanson R (2020) The elephant in the brain: Hidden motives in everyday life. Oxford University Press, Oxford
Smith A (1759) The theory of moral sentiments, edited by Raphael DD, Macfie, AL, 1982. Liberty Press, Indianapolis
Smith A (1776) An inquiry into the nature and causes of the wealth of nations, edited by Cannan E, 1976. The University of Chicago Press. Chicago
Stein Z (2019) Education in a time between worlds: Essays on the future of schools, technology, and society. Bright Alliance, London
Stiglitz J (2013) The price of inequality. W W Norton & Co Inc, London
Tocqueville A de (1835) Democracy in America, translated from the French origin by Lawrence G, 1990. Encyclopedia Britannica, INC, Chicago
Uzawa H (1965) Optimal technical change in an aggregative model of economic growth. International Economic Review 6:18–31
Von der Malsburg C (1997) The coherence definition of consciousness, in Proc IIAS Symp. on Cognition, Computation and Consciousness, 31 August–3 September, Kyoto, Oxford University Press
Ward M (1995) Butterflies and bifurcations: Can chaos theory contribute to our understanding of family systems? Journal of Marriage and Family 57: 629–689
Zhang WB (1991) Synergetic economics. Springer-Verlag, Heidelberg
Zhang WB (1992) A development model of developing economies with capital and knowledge accumulation. Journal of Economics 55:43–63
Zhang WB (2003b) Taiwan's modernization. World Scientific, Singapore
Zhang WB (2006) Growth with income and wealth distribution. Macmillan, London
Zhang WB (2008) International trade theory: Capital, knowledge, economic structure, money and prices over time and space. Springer, Berlin

Zhang WB (2021a) The butterfly effect in China's economic growth - From socialist penury towards Marx's progressive capitalism. Macmillan, London
Zhang WB (2021b) America the great and its self-destruction: Viewing the superpower through the lenses of ancient Confucianism and chaos theory. Algora Publisher, New York
Zinke FB (1868) Last winter in the United States. John Murray, London

Chapter 5
Uncertainties with Power, Wealth, and Sexuality

Spencer (1883) describes a relation between the social structure and slowly changed principles supporting the structure:

> Hence arises in the social organism, as in the individual organism, as the life of the whole quite unlike the lives of the units, though it is a life produced by them.... Systems are constituted by relatively permanent structures, existing in a relationship with their constituent units but not delimited in time by them: By a catastrophe, the life of the aggregate may be destroyed without immediately destroying the lives of all its units: while, on the other hand, if no catastrophe abridges it, the life of the aggregate immensely exceeds in the length the lives of its units (p. 479).

Complexity theory mathematically reveals how new structures will emerge in associations with interdependence among basic elements. History is full of new structures, bifurcations, catastrophes, and chaos. The world is now full of uncertainties because almost all parts of the world are closely connected with scientific knowledge, spread rationality, digital technologies, convenient and low-cost transportation systems, and global education. Slowly, but steadily converging global laws and rules, cultural exchanges, and (factually) free trade. At no time in history, mankind experiences such rapidly increasing complexity of human-made worlds originating from phantasies. Mankind has created high complexities that no person can fully understand about how they function. History shows that there is always a simple solution to man-made complexity: destroy the system and start a new epoch. China experienced such a collective catastrophe characterized by the ending of the Cultural Revolution not long ago. Historically, China has repeatedly destroyed the past dynasty and a new ignorant farmer generation re-constructed a new dynasty with increasing social complexity (to be destroyed again). The new generation with little formal education and complicated cultural experiences is full of energy with animal spirits (as Karl Marx and Keynes call them with different implications) and experiences positive growth in chaos before the system becomes corrupt and ineffective. It is also a global picture of empires in human evolution. The world is watching annually weakening processes of the superpower. Mankind is entering a kind of chaos and structural changes that mankind has not experienced yet. Emotion—as the yin of the

W. Zhang, *Complexity Theory and Uncertainties*, Understanding Complex Systems,
https://doi.org/10.1007/978-3-031-42394-9_5

organism—adds flavor to life (perhaps for reproduction) and rationality determines the trend of history. Man is born not for mere playfulness as emotional playfulness has no order. Healthy sustainability is a balanced yin–yang interdependence. Free playfulness leads to chaos with disorder and targeted rationality results in controlled order with boredom and destruction.

Stephen Hawking warns:

> Life on Earth is at the ever-increasing risk of being wiped out by a disaster, such as a sudden global nuclear war, a genetically engineered virus, or other dangers we have not yet thought of.

It might not be so surprising and harmful to Earth if the mother of mankind has only half of the current human population. Human selfish genes imply strong self-destructive mechanisms. Karl Marx argues:

> Men make their own history, but not make it as they please; they do not make under self-selected circumstances, but under circumstances existing already, given and transmitted from the past. The tradition of all dead generations weighs like a nightmare on the brains of the living. And just as they seem to be occupied with revolutionizing themselves and things, creating something that did not exist before, precisely in such epochs of revolutionary crisis they anxiously conjure up the spirits of the past to their service, borrowing from their names, battle slogans, and consumes in order to present this new scene in world history in time – honored disguise and borrowed language.

Man is a biological complexity equipped with spirit and knowledge. Man's spirit and thought are largely determined by time and space in a cultural sense. The single man does not show any significant improvement over many thousand years. Any man, if not influenced by any human culture from birth has a far higher probability to become the worst beast than to become a citizen in the standard of any modern society. Bourdieu (1992) expresses his approach to man and society:

> The effects engendered within fields are neither the purely additive sum of anarchical action nor the integrated outcome of a concerted plan... It is the structure of the game, and not a simple effect of mechanical aggregation (p. 17).

This reflects the spirit of a complex approach to exploring man and society. If man's growth process is an evolutionary process of complexity, traditional dividing and linearizing approaches to man can only lead to partial, if not biased, or even wrong descriptions of man. For instance, with the Newtonian approach, a medical doctor may have little knowledge about independence between mental health and physical health; a psychologist may not pay sufficient attention to the character formation of cultural, environmental, and economic forces on the customer; writers and poets may exaggerate emotions in human action; sociologists try to construct society even without proper understanding of its elements; historians believe in objective facts or data without theories or methods to check the credibility of objectivity; economists do not care about the role of emotions or irrationality on human action; moralists simply divide man into the moral man and the bad man without explicitly portraying infinity between the two, and philosophers use some simple elements in describing humanity without proper mechanisms of connecting those elements. It is rational to expect what

American thinker Thomas Sowell observes: "Facts do not speak for themselves. They speak for or against competing theories. Facts divorced from theories or visions are mere isolated curiosities." Each academic field does a wonderful job in their fields. But these fruits do not lead to an understanding of the whole when visualized in a more general vision of man and society (e.g., Byrne and Callaghan 2014).

Faced with the complexity of history and reality, Bourdieu (1998: 2) accepts the following approach:

> My entire scientific enterprise is indeed based on the belief that the logic of the social world can be grasped only if one plunges into the particularity of an empirical reality, historically located and dated, but which the objective of constructing it as our 'special case of what is possible action as Bachelard puts it, that is as an exemplary case in the finite world of possible configurations.

Globalization in association with digitalization and low-cost transportation implies that the logic of the social world can be hardly grasped by focusing on particular societies or civilizations because new structures, new faiths/ideologies, and new patterns of interactions will emerge. It does not require much talent to create a new ideology and does not need a deep heart to create a faith or religion for the masses to be addicted. But it is impossible to know the collective behavior of human masses. The study on well-ordered individual ant behavior and ant society illustrates the point.

What Thomas Jefferson says is still valid today: "He who knows nothing is closer to the truth than he whose mind is filled with falsehoods and errors." Modern people's mindsets are filled with fake news, fake histories, and rational theories built with narrow visions. The question is who cares about the truth? If one's mind is filled with falsehoods and errors and can still earn big profits, one probably does not care about the truth because one's nature may be born for profit rather than for truth. Honesty is a sure way to failure in modern society. One certainly becomes poor by being honest. Manipulation is the basic social skill for survival. Dishonesty may be an advantage in politics, business, and professionalism.

Life does not show a division of life, even society is seemingly characterized by an effective division of labor. Economics, politics, and other fields of social sciences should be logically connected. Intelligence is often trained to approach man and society by focusing on one aspect by assuming all other interactive parts as controllable "environments." Man can move toward self-destruction with efficient professionalism. China has a very efficient bureaucratic system equipped with talented and highly educated men over, at least, two thousand and five hundred years. One can easily see the consequence of this system in effectively controlling and manipulating mainland Chinese society. Complexity theory provides insights into why contemporary mainland China's political system emerges as a structure from other Chinese societies such as Taiwan, Singapore, and Hong Kong.

Karl Marx wraps up a relationship between human capital and technological changes: "The production of too many useful things results in too many useless people." Herbert Spencer first applies "survival of the fittest" to describe natural selection, after reading Darwin's *On the Origin of Species*, in his *Principles of Biology*.

Darwin later adopts Spencer's terminology to explore the implications of natural selection. For Darwin, survival of the fittest refers to better adaption for the immediate local environment, not a universally applicable index. In modern times, national performances are often measured in terms of GDP, the number of gold medals in international competition, freedom, openness, human rights, per capita GED, and the like. It is expected, theoretically, that those countries or races who have been displaying the excellence indexes in those commonly admired variables over the last half century are challenged with the possibility of becoming less influential, if not perish. There is an important index, at least from perspectives of biological reproduction and national or racial survival, called birth rate per woman in a lifetime. In some advanced economies, the rate is falling almost to 1, while it is necessary to have the rate over 2 for the given population group to replace itself over time. This falling tendency far below the reproduction rate occurs in those regions, which are characterized by free gender markets, high living standards, high education, and increased leisure time for cultivation and communications (within and between genders). Michel Foucault points out:

> If repression has indeed been the fundamental link between power, knowledge, and sexuality since the classical age, it stands to reason that we will not be able to free ourselves from it except at a considerable cost,

Cultivation, education, and virtue are essential for social order and economic development. Adam Smith does believe that the cultivation and virtue of common people are not key elements for an orderly prosperous society because the society will self-organize itself with the law and free market. Family, as a basic social organization, is not emphasized in Western social and economic thinking. In contrast, the family is the most essential institution in Confucianism. Cultivation and education begin at home, and the parent must see their children properly cultivated and educated. Although this belief was applied in Chinese civilization as the social, political, and economic mechanism to bring up children with the mindset to obey the authorities of all levels in society, the belief itself is to bring up children in a proper manner and with professional knowledge. Japanese society has made great efforts to continue this traditional way of thinking. Parents betray each other with sound reasoning of love for the sake of love and family institution is almost—rationally and structurally deterministically—broken in modern societies in a daily chaotic manner. Children begin to learn trust at home. It is not brutal forces in the family that make many children suffer in contemporary prosperous societies, but disloyalty and a divorced lifestyle that children have the first lesson of human trust and loyalty. Some contents of cultivation and education are transferred to schools and other educational institutions. When children make trouble, some parents tend to blame these institutions for not properly cultivating their children. One often sees shameless—typically rich and poor—parents and powerless educational institutions.

Politics cannot be independent of economic power and wealth distribution. In a market society like the United States, wealth is concentrated in a limited number of families. Rich businessmen control national media and can manipulate politicians so that national policies will guarantee the rich become, absolutely as well as relatively, richer. Few contemporary people can avoid economic problems in their lifetime. Economics deals with mundane issues. Almost any human networking and exchanges imply economics, even when people in action are not aware of them. Ludwig von Mises says:

> It is impossible to understand the history of economic thought if one does not pay attention to the fact that economics as such is a challenge to the conceit of those in power.

A politician is a bizarre career in a knowledge-based civilization because his professional advantage comes often from his ignorance about his professional questions. Modern top politicians tend to get used to talking about problems that they have neither concerns nor insights into and delivering promises even though they have little idea of how to carry them out. In billions of digital-connected screens, people across the Earth are watching simultaneously global politicians, some of them have neither intelligence, knowledge, education, nor morality to represent their countries. One might reasonably doubt human progress in politics over the last two centuries in light of extreme knowledge growth and technological changes.

Between birth and death, a man may experience all kinds of organic complexity, such as stationary states, smooth movements, bifurcations, structural changes, locking-in, catastrophe, and chaos. Immanuel Kant glorifies war:

> Even philosophers will praise war as ennobling mankind, forgetting the Greek who said: 'War is bad in that it begets more evils than it kills'.

The great German philosopher evaluates the positive sides of the war, while the negative sides are much lightened. Causes and consequences of war occupy most of the premodern historical writings of human evolution. In its civilizing process, mankind glorifies war as the royal road toward nobility. In modern times, one may die on battlefields even without the opportunity to show any personal character. British singer and composer Roger Waters (1943) calculates British benefits of colonizing the world:

> War is hugely profitable. It creates so much money because it's so easy to spend money very fast. There are huge fortunes to be made. So there is always an encouragement to promote war and keep it going, to make sure that we identify people who are 'others' whom we can legitimately make war upon.

American President Ronald Reagan told the real causes of many wars that America conducted: "History teaches that war begins when governments believe the price of aggression is cheap." War is often not conducted for some principle, faith, or nobility, but national (or rich class's and military business sector's) economic calculations. It is fuzzy to foresee wars in the future.

Samuel Johnson says: "Integrity without knowledge is weak and useless, and knowledge without integrity is dangerous." The current world has plenty of excellent experts in different fields who have integrity as highly paid professionals but even don't have opinions, for instance, on important issues, such as politics, trust, morality, and justice, for man and society. Unconnected professionals in society are good at applying—somehow localized—monopoly powers. Power is corrupt and absolute power is corrupt. Power is constructive or destructive but absolute power is destructive in the long term. Subtle corruption and deep destructiveness characterize modern societies. Mankind is facing great risks simply because no one knows what humans are collectively doing. Mankind is facing great risks unprecedentedly because humans have far greater destructive forces on themselves and the natural environment. Holland (1998: 3) portrays:

> I will restrict the study to systems for which we have useful descriptions in terms of rules or laws. Games, systems made up of well understood components.... are prime examples. Emergent phenomena also occur in domains for which we presently have few accepted rules.

There is no way to predict the paths of organic evolution. A prosperous and peaceful situation is not interpreted as a perfect and lasting happy state because the state will not last forever. The crisis is conceived as an opportunity for transformation. The term, *weiji,* which the Chinese use to paraphrase "crisis," is composed of the characters for "danger" and "opportunity." The term indicates a profound connection between crisis and change. To the Chinese mindset, history experiences cyclical processes of genesis, growth, breakdown, and disintegration. This dynamic vision is partly formed owing to the *Yi Jing* and partly owing to historical observations of traditional Chinese agricultural socioeconomic evolution. What Karl Popper says seems valid shortly:

> There is no history of mankind, there are only many histories of all kinds of aspects of human life. And one of these is the history of political power. This is elevated into the history of the world.

If mankind does not stay at the monkey level of collective evolution, mankind should do something interesting in politics to make mankind collectively wiser. Mankind naturally and rationally—certainly in terms of attractors in chaos theory—comes back to a stage of brutal competition if trust, morality, duty, sense of shame, and cultural cultivation do not interact properly. A man, a nation, or a race becomes a beast if he loses the sense of dutifulness, as Kant in the West and Confucius in East Asia claim. Let's be and be happy—it is what ancient philosopher Lao Zi advocates—under the condition that there were not many humans mingled together.

References

Bourdieu, P (1992) The logic of practice. Stanford University Press, Stanford. CA

Bourdieu P (1998) Practical reason: On the theory of action. Stanford University Press, Stanford, CA

Byrne D, Callaghan C (2014) Complexity theory and the social sciences: The state of the art. Routledge, London

Holland JH (1998) Emergence: From chaos to order. Oxford University Press, Oxford

Spencer H (1883) The principles of sociology, vol. 1. Appleton, London

Bibliography

Agnew TT (1978) Optimal exploitation of an employing a nonlinear harvesting function Ecological Modelling 6: 47–57

Haken H (1977) Synergetics: An introduction. Springer-Verlag, Berlin

Haken H (1983) The science of structure: Synergetics. Van Nostrand Reinhold

Laplace PS (1814 [1951]) A philosophical essay on probabilities. Dover, New York

Prigogine, I (1980) From being to becoming. WH Freeman,San Francisco

Prigogine, I (1997) The end of certainty Time, chaos, and the new laws of nature, written in collaboration with Stengers, I. The Free Press, New York

Zhang WB (2023a) Chaos, complexity, and nonlinear economic theory. World Scientific, Singapore

Zhang WB (2023b) Trust and justice: Complexity of man, complexity of society, and complexity theory. Lexington Books, Maryland

W. Zhang, *Complexity Theory and Uncertainties*, Understanding Complex Systems, https://doi.org/10.1007/978-3-031-42394-9

MIX
Papier aus verantwortungsvollen Quellen
Paper from responsible sources
FSC® C105338

If you have any concerns about our products,
you can contact us on
ProductSafety@springernature.com

In case Publisher is established outside the EU,
the EU authorized representative is:
Springer Nature Customer Service Center GmbH
Europaplatz 3, 69115 Heidelberg, Germany

Printed by Libri Plureos GmbH
in Hamburg, Germany